Remember Your Humanity:
Pathway to Sustainable Food Security

NIPA® GENX ELECTRONIC RESOURCES & SOLUTIONS P. LTD.
New Delhi-110 034

Remember Your Humanity:
Pathway to Sustainable Food Security

M.S. Swaminathan

NIPA® GENX ELECTRONIC RESOURCES & SOLUTIONS P. LTD.
New Delhi-110 034

NIPA® GENX ELECTRONIC
RESOURCES & SOLUTIONS P. LTD.

101,103, Vikas Surya Plaza, CU Block
L.S.C. Market, Pitam Pura, New Delhi-110 034
Ph : +91 11 27341616, 27341717, 27341718
E-mail: newindiapublishingagency@gmail.com
web: www.nipabooks.com

For customer assistance, please contact
Phone: + 91-11-27 34 17 17
Fax: + 91-11- 27 34 16 16
E-Mail: feedbacks@nipabooks.com

ISBN: 978-81-19254-34-7

Composed and Designed by NIPA®.

PREFACE

Many momentous events, leading to the achievement of seemingly impossible goals, have been the result of non-violent movements. Some examples are:

- Independence of India and Pakistan from colonial rule under the leadership of Mahatma Gandhi
- Civil Rights movement in the USA led by Martin Luther King, culminating in the election of Barack Obama as President of the USA
- End of apartheid in South Africa and the election of Nelson Mandela as President
- Breakdown of the Soviet Union led by Gorbachev and the fall of the Berlin Wall
- End of the Marcos dictatorship in the Philippines through the people power movement led by Corazon Aquino
- The Tahrir Square revolution in Egypt leading to the end of Mubarak's repressive rule

All these significant landmarks in human history and many more have been achieved through the Gandhian pathway of non-violence and Bertrand Russell and Albert Einstein plea that we should remember our humanity and forget everything else.

Let me also cite an example of happiness in one's life arising from tolerance and understanding. The world celebrated two years ago the bicentenary of Charles Darwin's birth and the 150th anniversary of the publication of the *Origin of Species*. Darwin's wife Emma Wedgwood was an ardent Catholic and did not believe in the evolution theory of her husband. Yet, they were a companionable couple and lived happily together. Emma even edited Darwin's book. When asked how she was living contentedly with a man whose scientific work was regarded by some as blasphemy, she said: 'Charles lives by reason, while I live by faith – we agree to disagree and this mutual respect for divergent views explains how we live happily together.' Such respect for pluralism and diversity is the need of the hour.

There is however a growing violence in the human heart and violent confrontations have become the method of settling disputes. Violence breeds violence. And hunger and poverty too breed violence. The UN Millennium Development Goals adopted by all Member States in the year 2000 represent a global common minimum programme for sustainable human security and well-being. The first among the 8 goals adopted for accomplishment by the year 2015 relates to reduction in the incidence of hunger and poverty. Unfortunately, recent reviews by FAO, IFPRI, World Bank and other agencies show that far from declining, hunger is increasing, particularly in South Asia and sub-Saharan Africa. FAO estimates that about 100 million more were added to the number of hungry persons during 2008-09, mainly as a result of rising food prices. It is also becoming evident that poverty is a major cause of hunger. Therefore, anti-poverty programmes have to accord priority to hunger elimination. The economic, ecological and social costs of hunger are high and hence this goal deserves to be on the top of the political agenda and public concern. We must use our intelligence and reason in humanistic endeavours to fight hunger.

This poem by W. H. Auden, one of my favourites (and the emphasis is mine), explains the significance of a coalition of the compassionate in helping to achieve the long cherished goal of a world without hunger.

Hunger allows no choice
To the citizen or the police.
We must love one another or die......
Defenceless under the night
Our world in stupor lies;
Yet, dotted everywhere,
ironic points of light
Flash out wherever the Just
Exchange their messages;

May I, composed like them
of Eros and of dust,
Beleaguered by the same
Negation and despair,
Show an affirming flame.

This volume is a collection of my thoughts and beliefs on a variety of subjects I have been deeply interested in and also equally deeply concerned with. It has been my constant refrain over the years, expressed in the many articles I have written and the talks I have given, that for lasting environmental security and human happiness, we need to work for a world devoid of both unsustainable lifestyles and unacceptable poverty. To face successfully the ecological, economic and social challenges confronting us today, we need transformational genes, technologies, and, above all, committed human beings.

Shri Sumit Pal Jain of New India Publishing Agency, leading publishers in the field of agriculture and allied sciences, has been urging me to send him some of my recent articles and talks to be brought out in a consolidated form. I am grateful to my colleague Dr. N. Parasuraman for taking the trouble to collate this collection and to my editor Ms. Gita Gopalkrishnan for suitably editing the material and putting it all together. I thank Mr. N. Ram, Editor-in-Chief, *The Hindu*, for permission to reproduce several of the articles which I had written for *The Hindu* over the years.

The papers in this book were written during the last three to four years. Some degree of duplication is therefore likely. They have been removed to the extent possible. I apologise for any avoidable duplication.

Chennai

14 November 2011

M.S. Swaminathan

CONTENTS

Preface v

Chapter-1 Remember Your Humanity 1

Chapter-2 Ethics and Science 7

Chapter-3 Four Pillars of Sustainable Human Happiness 13

Chapter-4 Norman Borlaug and His Fight against Hunger 19

Chapter-5 Bridging the Digital Divide: Empowering the People 25

Chapter-6 Harnessing the Demographic Dividend for Agricultural Rejuvenation 31

Chapter-7 Legislation for Food Security 35

Chapter-8 How to Resolve the Crisis of Indian Agriculture 39

Chapter-9 Gandhiji's Plea for a Hunger Free India: Current Reality and Way to Progress 45

Chapter-10 Wheat Imports and Food Security 53

Chapter-11 Distribute, Procure, Store and Sow – Steps to Food Security 57

Chapter-12 Sustainable Food Security: Pre-requisites for Success 63

Chapter-13 Pathway for Food Security for All 69

Chapter-14 Biodiversity and Poverty Alleviation 75

Chapter-15 Biodiversity and Sustainable Food Security ... 81

Chapter-16 Priorities in Agricultural Research and Education 91

Chapter-17 Managing Anticipated Food Crises 101

Chapter-18 To the Hungry, God is Bread 107

Chapter-19 The Wheat Mountains of the Punjab 113

Chapter-20 Fish for All and For Ever 119

Chapter-21 Leveraging Agriculture for Improving Nutrition and Health 125

Chapter-22 GM Food Crops: Risks and Benefits 129

Chapter-23 Land Rush and Food Security in an Era of Climate Change 133

Chapter-24 Darwinism in a Warming Planet 139

Chapter-25 Climate and Food Security 143

Chapter-26 Maximising the Benefit of a Good Monsoon 149

Chapter-27 Building Sustainable Water Security Systems . 153

Chapter-28 Drought Management for Rural Livelihood Security 157

Chapter-29 Privatisation of Food and Water Security Systems: An Unequal Social Bargain 161

Chapter-30 Shaping the Future of Agriculture in Northeastern India 167

Chapter-31 Towards Vidarbha's Agricultural Renaissance 173

Chapter-32 Science's Role in Stimulating West Bengal's Agricultural Performance 179

Chapter-33 Singur and Our Socio-Economic Future 187

Chapter-34 From Vision to Impact 193

Acronyms 205

Chapter 1

Remember Your Humanity

On August 6, 1945, the most dreadful among the weapons of mass destruction – the atom bomb – was dropped in the civilian area of Hiroshima. Three days later, another atom bomb was dropped in Nagasaki. In July 1955, Bertrand Russell and Albert Einstein issued their famous manifesto seeking the abolition of nuclear weapons and appealing to all inhabitants of Planet Earth:

> Remember your humanity, and forget the rest. If you can do so, the way is open to a new Paradise; if you cannot, there lies before you the risk of universal death.

In 1957, the Russell-Einstein Manifesto led to the birth of the Pugwash Conferences on Science and World Affairs, an organisation devoted to the causes of ending the nuclear peril and reminding scientists of their ethical responsibility for the consequences of their discoveries, particularly in the area of nuclear threat to human survival.

The Pugwash Conference held in 1995 at Hiroshima on the occasion of the 50th anniversary of the advent of atomic weapons concluded: 'The end of the Cold War, and the beginning of deep reduction in the huge nuclear arsenals that the war spawned, have provided an unprecedented opportunity for the abolition of nuclear weapons as well as the abolition of war.' Meeting again in Hiroshima in July 2005, the Pugwash Council observed:

> The decade since 1995, when Pugwash last met in Hiroshima, has been one of missed opportunities and a marked deterioration in global security, not least regarding the nuclear threat. In that time, additional states have acquired nuclear weapons, there has been little tangible progress in nuclear disarmament, new nuclear weapons are being proposed, and military doctrines are being revised that place a greater reliance on the potential use of such weapons.

The prospects for nuclear terrorism and adventurism have now become real. The voice of sanity of the survivors of the 1945 nuclear annihilation in Hiroshima and Nagasaki is yet to be heard. This is unfortunate since only they know what hell on earth means.

Members of the Pugwash Council, meeting just steps away from Hiroshima's ground zero, have hence appealed to fellow scientists and citizens to confront the threat of nuclear weapon use that could materialise at any time, without warning, in any part of the world. To political and government leaders, our message is simple, but stark: as long as nuclear weapons exist, they will one day be used.

The Seventh Review Conference of the Nuclear Non-Proliferation Treaty (NPT), held in the spring of 2005 in New York, ended in a deadlock. The five original nuclear-weapons states (US, Russia, UK, France and China) showed themselves unwilling to take decisive action to implement their obligations under Article VI of the NPT to move decisively toward the irreversible elimination of their nuclear arsenals. All states must share the blame for missing a solid opportunity at the Review Conference to resolve problems such as equitable access to civilian nuclear technologies, as allowed under Article IV, while at the same time tightening protections to ensure that such materials are not diverted to military use.

The broad framework of nuclear weapons disarmament is in danger of collapsing. The Comprehensive Test Ban Treaty (CTBT) has not entered into force, the US and Russia need to accelerate and enlarge the reductions called for by the Moscow Treaty, and negotiations have yet to begin on a Fissile Material Cut-off Treaty (FMCT) to eliminate production of weapons-grade highly enriched uranium (HEU) and plutonium. Far more needs to be done to control and dispose of existing stockpiles of HEU that run the risk of falling into the hands of terrorist groups. Large numbers of tactical nuclear weapons continue to be deployed in Europe and elsewhere, having no military rationale whatsoever, while pressures mount from certain quarters for developing and deploying space weapons.

The explosive progress in science and technology witnessed in recent decades has provided uncommon opportunities for realising the UN Millennium Development Goals in the areas of food, water, health, education and clean environment for all. Yet, most developing countries, including India, are falling behind the targets set. The extensive co-existence of unacceptable poverty and unsustainable lifestyles is not conducive to the creation of a climate for peace and harmony. What we urgently need is a shift in emphasis among militarily and economically powerful countries from military to moral leadership. At the same time, Einstein's advice to fellow scientists that 'concern for Man himself and his fate must always form the chief interest of all technical endeavours in order that the creation of our minds shall be a blessing and not a curse' should be the guiding motto in scientific laboratories everywhere in the world.

It will be useful to recall the role Jawaharlal Nehru played in mobilising scientific opinion against nuclear weapons. Early in 1954, he called for the setting up of a committee of scientists to explain to the world the effect a nuclear war would have on humanity. This idea was taken up by Joseph Rotblat, who along with Pugwash was awarded the Nobel Peace Prize in 1995, and Eugene Rabinowitch, resulting in the organisation of the Pugwash Conferences on Science and World Affairs. The name of the organisation comes from the Pugwash Village in Nova Scotia, Canada, where the first conference was held in 1957. Jawaharlal Nehru was also the first foreign Prime Minister to visit Hiroshima. In 1957, he praised the atom bomb survivors for their determination to spread around the globe information on the enormous harm that radiation can cause to both the present population and to the generations yet to be born. Even now, harmful mutations are being observed in children in Hiroshima and Nagasaki. Thus, the genetic harm is as serious as the immediate harm. Jawaharlal Nehru played a major part in getting the first UN Conference on the Peaceful Uses of Atomic Energy organised in Geneva in 1955. This conference was chaired by the late Dr. Homi Bhabha, the then Chairman of the Atomic Energy Commission, who outlined in his Presidential Address a strategy for harnessing the multiple contributions that nuclear tools can make to strengthen food, health and energy security in the world.

In my Presidential Address delivered at the Pugwash Conference held in Hiroshima on in July 2005, I outlined the following steps to achieve the goal of a nuclear- peril-free world.

- All nations with nuclear weapons should adopt a legally mandatory policy of "no first use of nuclear weapons", as homage to the survivors of the nuclear tragedy of 1945.
- Nations should respect commitments to the nuclear non-proliferation treaty (NPT), ratify the Comprehensive Test Ban Treaty (CTBT), conclude a Fissile Material Cut Off Treaty, and ban all research relating to the development of new nuclear weapons.
- A Nuclear Weapons Convention, outlining a road map for getting to zero by 2020, should be concluded.
- Prospects for nuclear terrorism and adventurism must be avoided by eliminating all unsecured nuclear fissile material and by implementing the concrete steps proposed by Pugwash for the elimination of highly enriched uranium; otherwise there is the risk of nuclear power groups and individuals emerging, in addition to nuclear power states.
- Because of the multi-dimensional threats posed to human security by climate change, and the consequent need for reducing greenhouse gas emissions, interest and investment in nuclear power plants are growing. The civilian uses of atomic energy are likely to grow. Hence, the UN may convene an International Conference on the Civilian Uses of Atomic Energy to develop a Code of Conduct to ensure that the non-military use of nuclear fuels does not get abused and to further strengthen safeguards and the inspection role and monitoring capacity of IAEA (International Atomic Energy Agency).
- Democratic systems of governance are fast spreading in the world, which involve the holding of free and fair elections periodically. It would be useful to develop a Hiroshima-Nagasaki 60th Anniversary Appeal which calls upon all political parties in every country to include in their next election manifesto, a firm commitment to work for speedy nuclear disarmament with a view to rid the world of the nuclear-peril as soon

as technically feasible. Without global political commitment, this goal cannot be achieved. At the same time, it would be useful to introduce in all school curricula information relating to the consequences of the use of nuclear weapons in Hiroshima and Nagasaki in August 1945, so as to bring home the immediate and long term disastrous impact of a nuclear war. Without public and political education, the climate for peace and nuclear disarmament will not exist.

Looking at the brighter side, nuclear weapons have existed for 60 years but have fortunately not been used. This is a tribute to the work of Pugwash and numerous civil society organisations. Unfortunately, the growing number of suicide bombing incidents indicates that we are now entering unchartered territory in human conflicts and retribution. At least to prevent the potential non-state use of nuclear weapons, nuclear weapon states should not lose even a day in working towards the goal of zero in the existence of such weapons.

❑❑❑

Chapter 2

Ethics and Science

An ancient Chinese proverb says

> If you are thinking one year ahead, plant rice
> If you are thinking ten years ahead, plant trees
> If you are thinking hundred years ahead, educate the people

Subramanya Bharati, a great Tamil poet, wrote decades ago that nutrition and education were the two legs of a human being — nutrition for the body and education for the mind. We must, therefore, foster a movement for integrating academic excellence and social relevance in the curricula of our educational institutions. Our universities should promote the growth of the science of ecotechnology, which is the product of the integration of traditional knowledge with frontier science. This will need a mind-set change in relation to the knowledge and wisdom of our tribal and rural families.

In the 1990s, I developed the Iwokrama Rain Forest Programme, which represents the world's largest adventure in the sustainable management of rainforests. The local Amerindian population welcomed me to Iwokrama with a song, which translates thus:

The sky is held up by the forest.
If the forest disappears, the sky.
Which is the roof of the world collapses.
Nature and man then perish together.

It is such wisdom and ecological prudence that we must recapture today, when we see all around us the spread of a greed revolution with reference to the exploitation of natural resources.

With the growing power of human beings in the fields of genetic modification and nano technology, there is need for greater attention to bioethics. There are many ethical considerations in medical biotechnology, including the area of reproductive cloning. However, just because there are problems, we should not condemn the technology. Every area of frontier technology, such as nuclear sciences, can be used or abused. This is why the inclusion of bioethics in the curriculum becomes important. Our aim in biological sciences should be the promotion of an era of biohappiness, based on the sustainable and equitable conversion of bioresources into jobs and income. Biohappiness and not bioterrorism should be the end result of our scientific endeavour.

Translational research is another area which is worthy of greater attention in our universities. Prime Minister Dr. Manmohan Singh has pointed out that while C. V. Raman won the Nobel Prize as far back as 1930 for the Raman Effect, most of the instruments available in India today using this principle are imported. Translational research will help to convert scientific findings into commercially viable technology. While science advances the frontiers of knowledge, it is technology that converts scientific knowledge into products and processes and thereby generates wealth. Translational research is particularly needed in our country in areas relevant to the rural professions, including agriculture, where the gap between scientific know-how and field level do-how is widening.

2010-2020 has been declared as the Decade of Innovation. The Union Minister for Human Resource Development, Kapil Sibal, has expressed the hope that India will become an Innovation Superpower by 2030. There are also frequent announcements about India becoming a Knowledge Superpower. Unfortunately, those who make such statements, including the

Knowledge Commission, seem to overlook the substrate conditions necessary for our young people to become masters of innovation. It has been reported that out of every 100 campus candidates interviewed by reputed companies, only 10 to 20 have been found suitable for employment. This points to the need for attention to the quality of education and also to the creation of opportunities for integrating theory and practice. Taking into account the uncommon opportunities provided by modern information communication technologies, we need to restructure and reform our teaching and training methods. We need a Pedagogic Revolution, a learning revolution, in the country.

The other essential pre-requisite for achieving the position of a Knowledge and Innovation Superpower is the opportunity for every newborn child to achieve its innate genetic potential for physical and mental development. Every fourth child born in our country is characterised by low birthweight (LBW) due to maternal and foetal undernutrition. Nearly 45 per cent of children under the age of five in the country are underweight. Such children suffer from many handicaps, including reduced cognitive abilities. To become an Innovation Superpower, we must first fight intellectual dwarfism caused by maternal and infant malnutrition. We must adopt a whole life- cycle approach in our plans for food for all and for ever. We must not deceive ourselves into believing that by establishing 14 Innovation Universities, we will become an Innovation Superpower. It is worthwhile recalling what J.R.D. Tata once said: 'I do not want India to become a superpower; I want it to be a happy country.' Nutrition and education are the pathways to a happy country.

Food and drinking water are the first among the hierarchical needs of a human being. Food security at the level of each individual child, woman and man is hence the first requirement for a healthy and productive life.

The three major components of sustainable food security are:

- Availability of food in the market, which is a function of internal production, and, where essential, imports.
- Access to food, which is a function of adequate purchasing power, and
- Absorption of food in the body, which is a function of clean drinking water, sanitation and primary healthcare.

The proposed National Food Security Act is being designed to ensure economic access to food through legal entitlement, while factors relating to food production and absorption are proposed to be included as essential enabling provisions. In this context, it may be worthwhile drawing attention to the unique structure of Indian agriculture, as compared to the role of farming in industrialised countries.

a. In industrialised countries, farming is a food or other commodity producing machine, while in our country, farming is the backbone of the livelihood security system for over 60 per cent of the population.

b. In industrialised countries, less than 3 per cent of the population is engaged in farming and may be called "farmer-consumers." However, in India, over 60 per cent of the population belong to the "farmer-consumer" category.

c. According to the 2001 census. 70 per cent of our population lived in rural areas. The 2011 census may probably show a rural-urban population ratio of 65:35, taking into account the growth of "rurban" areas, as in Kerala and the Punjab.

d. Our 80 per cent of the over 115 million farming families belong to the small (2 ha and less) and marginal (1 ha and less) categories. Around 60 per cent of the area cultivated depends only on rainfall for crop cultivation. Dry-land farmers cultivate climate-resilient crops like millets, pulses and oilseeds such as castor. There is widespread malnutrition in the families of small and marginal farmers, as well as among share-croppers, tenants and landless labour. Therefore, increasing the productivity, profitability and stability of small farms will, as a single step, make the largest contribution to overcoming endemic hunger caused by inadequate purchasing power.

Several genuine concerns have been expressed by a committee chaired by Dr. C. Rangarajan with reference to our capability to increase food production to the extent needed to fulfil legal entitlements, the impact of higher government procurement on open market prices, and the total cost of the subsidy involved.

These concerns can be addressed only by long-term policy changes in both agrarian reform and agricultural revitalisation, as set out in detail in both the Report of the National Commission on Farmers (2006) and the

National Farmers' Policy (2007). Quick-fix solutions like "packages" can only make a temporary difference. Since the major concerns of the Rangarajan Committee relate to production and procurement, I wish to address them briefly

Production: India has a vast untapped production reservoir in most farming systems, even with the currently available technologies. The gap between potential and actual yields ranges from 100-300 per cent in both rain-fed and irrigated areas, as per the data available with ICAR (Indian Council of Agricultural Research), Agricultural Universities, ICRISAT (International Crops Research Institute for the Semi-Arid Tropics) and Krishi Vigyan Kendras. A well planned "bridge the yield gap movement" on the lines proposed by the National Commission on Farmers will help to enhance the productivity and profitability of small holdings, leading to the alleviation of hunger among farm families as well as the other citizens needing nutritional support.

Procurement: Procurement stimulates and sustains production. Without procurement we would not have had the Green Revolution in the 1960s and 70s. This is not only true for our country, but even for the United States where the PL-480 procurement helped to sustain farmers' enthusiasm. In contrast, there has been no Green Revolution in Africa in spite of the enormous amount of money invested. This is because prices collapse when production goes up. Most African nations do not have a machinery to purchase grains at a minimum support price announced at the time of sowing. The presence of such a mechanism is our strength.

We should therefore emphasise that what the government should do is to intensify efforts in production and procurement and not continue the status quo with reference to millions of our children and countrymen remaining hungry. The following should be some of the essential components of a National Food Security Act.

1. Adopt a lifecycle approach, to legal entitlements, starting with pregnant mothers. A "First 1000-Days Child Nutrition and Development Programme" should be organised to provide nutritional support to pregnant women so that the newborn child has an opportunity for the full expression of its innate intellectual potential.

2. Enlarge the food basket to include nutritious millets (*bajra, jowar, ragi,* maize and minor millets) in the Public Distribution System, thereby

achieving double benefits, namely, improving nutrition security, while at the same time providing a market for the crops of dryland farmers and tribal families. Over 10 million tonnes of these crops can be procured from dry-farming areas, much of it from tribal families. These crops are also capable to some extent of withstanding drought and adverse climatic factors. They will therefore help to promote climate-resilient agriculture, an important need of the future.

3. Develop a decentralised procurement system and a national grid of community Grain Banks, rural godowns and modern storage structures.

4. Under enabling provisions, the highest priority should go to increasing agricultural productivity, so as to meet the food requirements of 1.2 billion human population and 1 billion farm animals. In addition, the effective implementation of the Rajiv Gandhi Drinking Water Mission, the Total Sanitation Programme, and the National Rural Health Mission needs high priority. Also, nutrition considerations should be mainstreamed into both the National Horticulture Mission and the Food Security Mission, in order to overcome hidden hunger caused by the deficiency of micronutrients like iron, iodine, zinc, vitamin A and vitamin $B_{12.}$ Horticultural remedies should be promoted for overcoming nutritional maladies.

5. In the case of diseases like HIV/AIDS, tuberculosis and leprosy where prolonged treatment is necessary, a food-cum-drug approach should be adopted, since undernutrition reduces the efficacy of the drugs prescribed.

In the Russell-Einstein Manifesto of 1955 we find the following said in the context of potential nuclear conflicts.

> There lies before us, if we choose, continual progress in happiness, knowledge and wisdom. Shall we, instead, choose death, because we cannot forget our quarrels?

We must remember our humanity and cultivate the culture of compassion, so that we can lead fulfilling lives.

❑❑❑

Chapter 3

Four Pillars of Sustainable Human Happiness

Bhutan has developed the concept of Gross National Happiness (GNH) as a substitute for Gross Domestic Product (GDP). In this theory, GNH comprises spiritual and cultural values, including love of sport, music, dance and spiritual activities. This is a good departure from measuring happiness purely from the view point of money. I would like to elaborate on four pillars that are necessary to uphold sustainable human happiness.

Ecology

There is an increasing awareness of the need for adopting sustainable lifestyles which will not make a heavy demand on our life-support systems of land, water, biodiversity, forests, oceans and climate. In his book titled *World on the Edge*.[1] Lester Brown has described how we still have time to prevent environmental and economic collapse. Recent calculations on the ecological footprint each one of us is leaving behind shows that our per capita consumption of natural resources is exceeding the biocapacity of our earth. Soon we will need two planets to meet the growing demand for land, water, forests and biodiversity. In this context, we should keep in view what Mahatma Gandhi said long ago: 'Nature provides for everyone's needs, but not for

[1] Brown, Lester R. (2011). *World on the Edge*. New York and London: W.W. Norton

everyone's greed.' We should prevent the spread of a greed revolution and promote an ever-green revolution which can help us to improve farm productivity in perpetuity without associated ecological harm.

The March 2011 earthquake in Japan, measuring 9.0 on the Richter scale, shifted the Honshu island by 2.4 metres. It also led to a 23 foot-high tsunami. Compounding these two disasters has been the damage caused to several atomic energy reactors in this region. The Fukushima Daiichi Atomic Power Plant has been particularly severely affected. Radiation threats are persisting. In spite of all our technological advances, we cannot easily overcome nature's fury. The calamitous events in northeastern Japan have however revealed the power of social engineering and cohesion in fostering a "we shall overcome" spirit. Ms. Nobuko Horibe of UNFPA (United Nations Population Fund), for example, has pointed out that the Japanese strength lies in their education system. To quote her: 'From kindergarten to elementary school and onward, the performance of students is measured by how a group performs. If one group is better than the others, members of that group are expected to help the weaker ones, so that no one is left out.' The Japanese system of education leads to caring for others and to selflessness, and thereby helps to foster high social synergy and promotes a coalition of the compassionate in times of distress. The Japanese education system also promotes inquiry-based instruction that activates students' curiosity in exploring how the world works. This is an asset during calamities.

As I was watching the threats to the nuclear power plants caused by the titanic tsunami in Fukushima, my memory went back to 1989 when as the then President of IUCN (International Union for the Conservation of Nature and Natural Resources), I had a discussion with Japanese scientists on the regeneration of mangroves along the coasts of Japan. Some scientists belonging to the older generation mentioned the beneficial role played by mangroves in reducing the fury of coastal storms and tsunamis, as they serve as speed breakers. We then decided to establish, with the help of UNESCO, an International Society of Mangrove Ecosystems (ISME) in Okinawa, where once there were dense mangrove forests. I was the Founder President of ISME and promoted the preparation of a Charter for Mangroves.

During the tsunami which affected Tamil Nadu and other southern states in December 2004, the coastal communities observed that dense mangrove forests served as bio-shields, reducing the damage done. We therefore launched a programme both in India and Sri Lanka to plant mangrove and non-mangrove bio-shields.

The concern now about the safety of nuclear power plants located along the coast such as Kalpakkam and Kudankulam makes me feel that in addition to steps related directly to the design of the nuclear power stations, we should promote bioshields comprising mangrove and non-mangrove species in the coastal areas adjoining nuclear power plants. For this purpose, it may be worthwhile declaring such areas as Critically Vulnerable Coastal Areas (CVCA).

Equity

Equity is now considered both with reference to intra-generational effects and inter-generational consequences. For example, the most serious form of intra-generational equity is maternal and foetal undernutrition resulting in the birth of children characterised by low birthweight. As I have repeatedly pointed out, such LBW children suffer from many handicaps in later life, including impaired cognitive abilities. Thus, even at birth they are denied opportunities for realising their innate genetic potential for physical and mental development. Unfortunately, nearly every fourth child born in our country has a low birthweight because of poverty at the household level. Over forty years ago, the Government of India introduced an imaginative programme, the Integrated Child Development Services (ICDS). This programme was to meet the nutritional, health and educational needs of the child in an integrated manner. In spite of ICDS, we have serious problems of malnutrition, stunting, wasting and other abnormalities among children. We should now redesign ICDS and deliver the services in two phases of a child's life. First, we should initiate a 1000-Days Child Development Service programme which will begin with conception and extend up to two years. Such a programme will need to pay attention to the nutrition security of the mother during the pregnancy period and later to both mother and child.

Ethics

Again, a point I have mentioned previously is that there is need for greater attention to bioethics in the fields of genetic modification and nano technology. Notwithstanding the many ethical considerations in areas of biotechnology that ought to be addressed, we should not condemn the technology and throw the baby out with the bath water. Every area of frontier technology, such as nuclear sciences, can be used or abused. This is why the inclusion of bioethics in the curriculum becomes important. Our aim in biological sciences should be the promotion of an era of biohappiness, based on the sustainable and equitable conversion of bioresources into jobs and income.

Another area of ethics needing attention is the restoration of the reverence paid to foodgrains which has been part of Indian culture right from the Vedic period. Thanks to the hard work of our farm women and men and the favourable weather, the wheat harvest in 2011 may exceed 82 million tonnes. The Central and State governments may have to purchase, distribute and store over 26 million tonnes of wheat during April-June 2011. Harvesting of the world-famous Malwi wheat has already started in Madhya Pradesh. Over 47 million tonnes of wheat and rice are currently in government storage facilities. It is reported in the media that out of the 15 million tonnes of additional storage facility for which funds have been sanctioned, hardly one per cent of the target has been achieved. In the *Upanishads*, foodgrains were regarded sacred and the wastage of grains a serious sin. Invariably, there are rains and even hailstorms in the Punjab-Haryana-western Uttar Pradesh region during April-May. Moisture is the greatest enemy of food safety, since it helps the development of mycotoxins. Therefore, there is need for a post-harvest management strategy which can safeguard the harvested grains from both the quantitative and qualitative perspective.

Economics

Financial balance sheets generally determine our growth rate. We have had an impressive growth rate in GDP in recent years, in spite of the inadequate progress of the farm sector. According to the Economic Survey 2011, the contribution of agriculture and allied sectors to GDP has come down below 15 per cent. Nevertheless, the onus of employment is largely on the agriculture and allied sectors. This explains why there is persistence of unacceptable levels of poverty in our country. Fortunately, several steps have been taken recently to revive our agricultural progress.

The other important recent development is the shifting of a patronage approach to a rights approach. For example, we now have legal guarantees for information, education, employment and land rights to tribal families.

Thus, there is a welcome shift from patronage to rights in relation to basic human needs. There is currently a commitment to include the right to food also among the legal rights. The National Food Security Bill based on a lifecycle approach to the right to food, when finalised and enacted by Parliament, will become the brightest jewel in the crown of Indian democracy.

High economic growth rates and extensive deprivation co-exist in our country. The way in which we can all contribute to ending this dichotomy will be to follow the advice of Swami Vivekananda:

> This life is short,
> Its vanities are transient
> He alone lives who lives for others.

❑❑❑

Chapter 4

Norman Borlaug and His Fight against Hunger

Norman Borlaug's early upbringing in an Iowa farm and his experience of hardship during the US Depression of the early 1930s instilled in him the desire to bring science to address the problems of low farm productivity, poverty and hunger. After completing his Ph.D degree in Plant Pathology at the University of Minnesota in 1942, he joined the Rockefeller Foundation's agricultural programme in Mexico, leading to the birth of the International Maize and Wheat Improvement Center (CIMMYT). There he began his work on wheat, concentrating on the control of stem, stripe and leaf rusts, the important diseases of the crop. He introduced a multi-pronged approach to manage the rust fungus, including the development of composite varieties characterised by phonotypic identity but genotypic diversity with reference to resistance to different races of the pathogen, an approach conferring enduring resistance as a result of reduced pressure on the fungus to mutate and create more virulent strains.

Borlaug started his research career in agriculture in Mexico at a time when the world was passing through a serious food crisis. During 1942-43, nearly 2 million children, women and men died of hunger during the great Bengal Famine. China also experienced widespread and severe famine during the 1950s. Famines were frequent in Ethiopia, the Sahelian region of Africa and many other parts of the developing world. It is in this background that Borlaug decided to look for a permanent solution to recurrent famines by harnessing science to increase the productivity, profitability and sustainability of small farms.

Drawing on the availability of the Norin 10 dwarfing gene from Japan after World War II, Borlaug launched a programme to breed semi-dwarf high-yielding varieties of wheat, which responded well to irrigation and fertiliser application. Traditional wheat varieties were tall and hence would lodge, if grown under high soil fertility conditions. By the late 1950s, Borlaug had developed several semi-dwarf spring wheat varieties, capable of yielding 5 to 6 tonnes per hectare. Since conditions for good crop growth are also conducive to the spread of pathogens, Borlaug intensified his research on combining high yield potential with a high degree of resistance to the major diseases, particularly rusts, through gene pyramiding.

He carried forward this battle against virulent races of wheat rusts to his dying day. Worried about the rapid spread of a new race of stem rust (*Puccinia graminis*) from Uganda, named UG 99, he organised, in 2007, a Global Rust Initiative to check the spread of virulent strains like UG 99 through basic research, surveillance and breeding. In March 2009, at the age of 95, Borlaug assembled over 300 wheat breeders and pathologists from around the world at Ciudad Obregon, Mexico, and told them: 'There is no room for complacency. Let us get on with the job of eliminating the rust menace.'

When he initiated the semi-dwarf wheat breeding programme in 1953, Borlaug decided to adopt a shuttle breeding programme involving the growing of different segregating generations like F_2, F_3, etc., under two diverse growing conditions — a summer crop in the cooler highlands near Mexico City and a winter crop in the warm weather conditions prevailing in Sonora. Such alternate selection under two different temperature and daylight conditions led to the breeding of semi-dwarf wheat strains with broad adaptation like Sonora 63, Sonora 64, Lerma Rojo 64, Mayo 64, etc. These were the varieties combining a high yield potential (5 to 6 t/ha) with resistance to rusts and wide adaptation which became the catalysts of the wheat revolution witnessed in Mexico, India and Pakistan in the 1960s.

In 1966, India imported 18,000 tonnes of seeds of Lerma Roja 64 A and a few other varieties from Mexico with the help of Borlaug, as a part of a "purchase time" strategy, resulting in a quantum jump in wheat production from 12 million tonnes in 1965 to 17 million tonnes in 1968. Similar results were being obtained in rice, as a result of the introduction of the Dee-gee-woo-gen dwarfing gene from China in tall varieties of *indica* rice at the International Rice Research Institute in the Philippines. Dr William Gaud of USA coined the term 'Green Revolution' in 1968 to denote productivity-led advances in production. For example, India produced 80 million tonnes of wheat from 26 million ha in 2009. If this production was to be achieved at the pre-Green Revolution yield level of 1 t/ha, 80 million hectares would have been needed. This is why the Green Revolution is also referred to land- or forest-saving agriculture.

The catalyst of the miracle was the new plant type sent by Norman Borlaug in 1963. This plant type had a semi-dwarf plant stature and was capable of utilising fertiliser and water very efficiently. When grown with good agronomic practices and soil fertility management, varieties like Lerma Rojo 64A and Sonora 64 gave about 5 tonnes of wheat per hectare, in contrast to 1 to 2 tonnes per hectare of the earlier tall varieties. The earlier varieties like C306 bred by Chowdhry Ramdan Singh had amber grains and excellent *chapathi*-making properties. Fortunately Borlaug had also sent segregating populations from which wheat breeders at the Indian Agricultural Research Institute, New Delhi and the Punjab Agriculture University, Ludhiana selected high-yielding amber grain and good culinary quality varieties like Kalyan Sona and Sonalika. This resulted in enormous enthusiasm among the farmers of the Punjab, Haryana and western Uttar Pradesh and I described the role of farmers in the revolution as follows:

> Brimming with enthusiasm, hard-working, skilled and determined, the Punjab farmer has been the backbone of the revolution. Revolutions are usually associated with the young, but in this revolution, age has been no obstacle to participation. Farmers, young and old, educated and uneducated, have easily taken to the new agronomy. It has been heart-warming to see young college graduates, retired officials, ex-army men, illiterate peasants and small farmers queuing up to get the new seeds. At least in the Punjab, the divorce between intellect and labour, which has been the bane of our agriculture, is vanishing.

Having made a significant contribution to shaping the agricultural destiny of many countries in Asia and Latin America, Borlaug turned his attention to Africa in 1985. With support from President Jimmy Carter, the late Ryoichi Sasakawa, Mr Yohei Sasakawa and the Nippon Foundation, he organised the Sasakawa-Global 2000 programme. Numerous small-scale farmers were helped to double and triple the yield of maize, rice, sorghum, millet, wheat, cassava and grain legumes. Unfortunately, such spectacular results in demonstration plots did not lead to significant production gains at the national level, due to lack of infrastructure such as irrigation, roads, seed production and remunerative marketing systems. The blend of professional skill, political action and farmers' enthusiasm needed to ignite a Green Revolution as in India was lacking in Africa at that time. This made him exclaim, 'Africa has the potential for a Green Revolution, but you cannot eat potential.'

Concerned with the lack of adequate recognition for the contributions of farm and food scientists, Borlaug had the World Food Prize established in 1986, which he hoped would come to be regarded as the Nobel Prize for food and agriculture. My research centre in Chennai in India is the child of the first World Food Prize I received in 1987. Throughout his professional career, Borlaug spent time in training young scholars and researchers. This led him to promote the World Food Prize Youth Institute and its programme to help high school students work in other countries in order to widen their understanding of the human condition. This usually became a life-changing experience for them.

The five principles Borlaug adopted, in his life, to quote his own words, were:

- Give your best.
- Believe you can succeed.
- Face adversity squarely.
- Be confident you will find the answers when problems arise.
- Then go out and win some bouts.

These principles have shaped the attitude and action of thousands of young farm scientists across the world. He applied these principles in the field of science and agricultural development, but I guess he developed them much earlier in the field of wrestling, judging from his induction into the Iowa Wrestling Hall of Fame in 2004.

I was privileged to be present, when he was awarded the Congressional Gold Medal in 2007. On that occasion he pointed out that between the years 1960 and 2000 the proportion of the world's people who felt hunger during some portion of the year had fallen from perhaps 60 per cent to about 14 per cent. The latter figure, he went on, still 'translates to 850 million men, women and children who lack sufficient calories and protein to grow strong and healthy bodies.' He added: 'The battle to ensure food security for hundreds of millions of miserably poor people is far from won.' This is the unfinished task that Norman Borlaug leaves to scientists and political leaders worldwide.

In the Indian spiritual text *Bhagavad Gita*, there is a saying that the Divine manifests itself in various forms, whenever there is acute suffering or injustice on Earth. I feel it will be appropriate to consider Borlaug as one such messenger who came to the rescue of those struggling for their daily bread. I am saying this because Dr. Borlaug was not only a great scientist but also a humanist full of compassion and love for fellow human beings, irrespective of race, religion, colour or political belief. This is clear from his last spoken words on the night of Saturday 12 September 2009. Earlier in the day, a scientist had shown him a nitrogen tracer developed for measuring soil fertility. His last words were:'Take the tracer to the farmer.' This life-long dedication to taking scientific innovation to farmers without delay set Borlaug apart from most other farm scientists carrying out equally important research.

When Mahatma Gandhi died in January 1948, Prime Minister Jawaharlal Nehru said:

> The light has gone out of our lives.....the light that shone in this country was no ordinary light....a thousand years later, that light will be seen in this country, the world will see it and it will give solace to innumerable hearts. For that light... represented the living, the eternal truths, reminding us of the right path, drawing us from error, taking this ancient country to freedom.

The same can be said of Norman Borlaug, who strove to take humankind to freedom from hunger and deprivation. His repeated message that there was no time to relax until hunger became history will be heard so long as a single person is denied opportunity for a healthy and productive life because of malnutrition.

❑❑❑

Chapter 5

Bridging the Digital Divide: Empowering the People

The country is at long last becoming sensitive to the serious consequences of the growing rural-urban divide in terms of investment, infrastructure and opportunities for income and employment. The rural-urban divide also leads to an expanding rich-poor divide. Since crop and animal husbandry, fisheries, forestry, and agro-processing are the main sources of rural livelihoods, the current agrarian crisis is adding to the problems of hunger, poverty and unemployment. According to the Union Planning Commission, we are off-track in achieving most of the UN Millennium Development Goals. A major cause for the growing rich-poor divide both between and within nations is unequal access to modern technology. Technology helps to achieve a paradigm shift from unskilled to skilled work and thereby move large numbers of the rural poor from the primary to the secondary and tertiary sectors of economic activity. If technology has been a major factor in promoting economic and social divides in the past, the challenge now lies in enlisting technology as an ally in the movement for economic, gender and social equity.

Keeping the above in view, the M.S. Swaminathan Research Foundation (MSSRF) has been working during the last 15 years on the skill and knowledge empowerment of the rural poor based on a pro-nature, pro-poor and pro-woman orientation to technology development and dissemination. Further, the aim has been to substitute jobless economic growth with job-led growth. This is particularly essential in the present context when unemployment and under-employment are taking a heavy toll on the morale of the youth. Similarly, "technology fatigue" in agriculture caused by inadequacies in research and extension efforts has led to increasing indebtedness among farm households. Although there has been much effort to increase and streamline institutional credit, small farmers still depend upon moneylenders for a variety of reasons. Farm women have by and large been bypassed by the institutional credit system, since they do not have ownership rights over land.

Modern agriculture is becoming knowledge intensive. Farmers need both generic and dynamic information on matters relating to farm operations and markets. The extension system by and large has not been able to respond to their needs, particularly in the area of dynamic information and advice on economically viable crop diversification and land and water use based on meteorological and marketing factors. Trade, quality and genetic literacy is low both among farm and fisher communities.

The work undertaken by MSSRF in setting up community-centered and -managed Village Knowledge Centres (VKCs) based on modern information and communication technologies (ICT) has shown that ICT helps to improve the timeliness and efficiency of farm operations and enhances income through producer-oriented markets. Also, experience has shown that bridging the digital divide is a powerful method of bridging the gender divide. Knowledge connectivity therefore confers multiple economic and social benefits. VKCs operate on the principle of social inclusion. The information provided, which includes location-specific data on entitlements to different government schemes, is demand driven and is in the local language. For example, in Pondicherry there are over 150 government schemes designed to help the poor; yet nearly 20 per cent of families are below the poverty line. After the onset of the digital age, knowledge on entitlements and how to access them has grown rapidly.

Encouraged by the ability of rural women and men to take to ICT like fish to water, the MSSRF initiated two major steps in 1993, to take ICT to every one of the over 600,000 villages in India. The first is the organisation of a National Alliance for Mission 2007: Every Village a Knowledge Centre. The second is the establishment of the Jamsetji Tata National Virtual Academy for Rural Prosperity.

Every Village a Knowledge Centre

The aim of this programme is to bridge the urban-rural digital divide and to harness ICT for addressing the major problems of rural India like poverty, illiteracy, ill-health, and low farm productivity. Thanks to widespread support for this goal, a National Alliance for Mission 2007 was formed in 2003. This Alliance has grown and now consists of over 400 members drawn from the public, private, academic, civil society and financial sectors. It currently includes 22 government organisations including the Department of Information Technology, the Ministry of Panchayati Raj, the Telecom Regulatory Authority of India, and Bharat Sanchar Nigam Limited; 94 civil society organisations; and 34 private sector information and communication technology (ICT) leaders such as NASSCOM (National Association of Software and Services Companies), TCS, HCL and Microsoft. Besides, 18 academic institutions such as the Indian Institutes of Technology and the Indira Gandhi National Open University and 10 financial institutions such as the National Bank for Agriculture and Rural Development (NABARD) and the State Bank of India are involved. The Alliance is supported by an international support group of UN and bilateral agencies and private sector corporations. This goes to prove that seemingly impossible tasks can be achieved by mobilising the power of partnership, since irrespective of the individual strengths of the alliance partners their collective strength becomes considerable.

From the beginning the Alliance partners have been committed to taking ICT to rural India on the principle of social inclusion in access to this valuable technology. Concurrent attention has been paid to connectivity, content creation, capacity building, care and management and coordination between knowledge and its field application, i.e., bridging the know-how / do-how gap.

The VKC is based on the principle of an integrated and appropriate use of the internet, cable TV, cell phone, community radio, and the vernacular press. To begin with, VKCs will be established in the 240,000 panchayats and local bodies. With the help of loud speakers and FM radio, they will be able to cover all the 600,000 villages in the country. internet-community radio and cell phone-community radio are powerful combinations for reaching the unreached with timely information. The spot prices of agricultural commodities monitored on a cell phone can be communicated to several villages through a low power FM radio. A group of VKCs will be supported by a block-level Village Resource Centre (VRC), which will also provide tele-conferencing facilities.

Apart from a sense of ownership by local women and men, the other major requirements for the success of the Village Knowledge Centre movement are in the areas of connectivity, content, and capacity building. Fortunately, the more than 670,000 km of buried fibre optic cable network nationwide offer the capability to connect an estimated 85 per cent of the villages.

The relevance and timeliness of the content will determine the interest of rural families in VKCs. The content should be demand-driven and area-, culture-, and time-specific. The National Alliance has suggested that at the level of each district a content consortium may be organised to enable VKC managers to access the information they need in the area of weather, health, entitlements to government projects, e-governance, credit and insurance, agriculture and market. The managers of VKCs will have to maintain active contact with the content consortium. To facilitate the availability of the right information at the right time and place, it is proposed to organise national digital gateways for agriculture, education, health and livelihoods. When the Alliance completes its mission, India will be the first developing country where the power and opportunity provided by ICT will reach every home and hut.

Jamsetji Tata National Virtual Academy

Besides connectivity and content, capacity building is essential for ensuring local ownership of VKCs. This is where the Jamsetji Tata National Virtual Academy (NVA) of MSSRF hopes to play a key role. To accord social prestige to grass-root ICT workers, a Jamsetji Tata National Virtual Academy for Rural Prosperity was established in 2004 with generous support from the Tata Trusts. This Academy has now over 1000 Fellows from all parts of the country and from a few neighbouring countries. The grass-root academicians feel a sense of pride in belonging to the Academy and have become the torch bearers of the rural knowledge revolution. In the words of President Dr. Abdul Kalam, the Academy represents 'the celebration of India's rural core competence'.

NVA trains at least one woman and one man from each village in computer literacy. Those who reveal the capacity to become master trainers are elected Fellows of the NVA. The Fellows of NVA are rural women and men who have studied up to the tenth class or up to the first degree. They will be master trainers and undertake the training of other rural women and men as well as children. The experience gained under MSSRF's VKC programme during the last eight years shows that rural women in particular are able to master ICT within a fortnight provided the pedagogic methodology is learning by doing.

With the help of the Indian Space Research Organisation (ISRO), additional centres are being opened in tsunami-affected areas and in farmers' distress hot spots in Kerala, Andhra Pradesh, Maharashtra, and Karnataka. These are areas where suicides by farmers occur. Those operating the computer-aided knowledge system at such centres will be either wives or daughters or sons of those who were driven to take their lives. This will help to provide a sense of realism and urgency in achieving a match between content and the need to save livelihoods and lives.

In 2007 the Alliance partners decided to continue this movement under the name Grameen Gyan Abhiyan (Rural Knowledge Movement). In order to build the capacity of both the Fellows of the Academy as well as other grass-root ICT workers in villages, a Jamsetji Tata Training School was established in 2006. The Alliance has developed the following strategy for taking the benefits of ICT to all parts of the country:

- Block Level: Village Resource Centres (VRCs) established with the help of ISRO. VRCs have satellite connectivity and tele-conferencing facilities in order to provide e-health, e-literacy, e-commerce, and other demand-driven and dynamic services.
- Panchayat / Local Body: Village Knowledge Centres or Gyan Chaupals with internet connectivity and with training facilities for local village school students
- Last Mile and Last Person Connectivity: This will be achieved through the integrated use of the internet and cellular phones or FM radio or public address systems.

We feel confident that all this can be achieved since the Government of India has included rural knowledge connectivity under its visionary Bharat

Nirman Programme and has started setting up Community Service Centres (CSCs) in 100,000 villages. In addition private sector companies like ITC will cover over 50,000 villages under the e-Chaupal programme. There are numerous other initiatives in the country sponsored by State governments, civil society organisations, academic institutions, as well as bilateral and multilateral agencies.

Jamsetji Tata showed how a combination of intellect and labour could help to make the impossible possible. By mobilising the power of partnership through the National Alliance for Mission 2007 and by harnessing the talent and commitment of the Fellows of the NVA, the goal of empowering rural families with the right information at the right time and place can be accomplished.

The Tunis World Summit on the Information Society held in 2005 demonstrated the spectacular progress made in technology development in just two years since the 2003 Geneva Summit. The world is thus witnessing two opposite trends. The explosive progress in science and technology is providing uncommon opportunities for health, food, water, work, energy, and literacy for all. On the other hand, a considerable proportion of humankind living under conditions of poverty, hunger, and deprivation feel a sense of social exclusion and injustice.

Consequently, there is a growing violence in the human heart. In the midst of a feeling of a brave new world of technological breakthroughs, the main news in the media every day is the loss of innocent lives caused by bomb explosions in different parts of the world. The extensive co-existence of unsustainable lifestyles and unacceptable poverty is not conducive to either harmony with nature or with each other. This is why bridging the digital divide is so important for human security and well-being in our country.

❑❑❑

Chapter 6

Harnessing the Demographic Dividend for Agricultural Rejuvenation

During his visit to India, President Barak Obama pointed out that our country is fortunate to have a youthful population with over half of the total population of 1.2 billion being under the age of 30. Out of the 600 million young persons, over 60 per cent live in villages. Most of them are educated. Gandhiji considered the migration of educated youth from villages to towns and cities as the most serious form of brain drain adversely affecting rural India's development. He therefore stressed that we should take steps to end the divorce between intellect and labour in rural professions.

The National Commission on Farmers (NCF)(2004-06) emphasised the need for attracting and retaining educated youth in farming. The National Policy for Farmers, placed in Parliament in November 2007, includes the following goal of the new policy:

> To introduce measures which can help to attract and retain youth in farming and processing of farm products for higher value addition, by making farming intellectually stimulating and economically rewarding.

At present, we are deriving very little demographic dividend in agriculture. On the other hand, the pressure of population on land is increasing and the average size of a farmholding is going down to less than 1 hectare. Farmers are getting indebted and the temptation to sell prime farmland for non-farm purposes is growing, in view of the steep rise in the price of land. Over 45 per cent of farmers interviewed by the National Sample Survey Organisation (NSSO) want to quit farming. Under these conditions, how are we going to persuade educated youth, including farm graduates, to stay in villages and take to agriculture as a profession ? How can youth earn a decent living in villages and help to shape the future of our agriculture ?

This will require a three-pronged strategy.

- Improve the productivity and profitability of small holdings through appropriate land use policies, technologies and market linkages, developing for this purpose a "4C approach", — conservation, cultivation, consumption and commerce.
- Enlarge the scope for the growth of agro-processing, agro-industries and agri-business and establish a "Farm to Home" chain in production, processing and marketing.
- Improve the productivity and profitability of small holdings through appropriate technologies and market linkages.

Opportunities in the services sector in rural India are crying for attention. Hence, I shall concentrate on giving a glimpse of the untapped opportunities awaiting our educated youth to take to a career of remunerative self-employment in villages. NCF had recommended a re-orientation in the pedagogic methodologies adopted in our Farm Universities, in order to make every scholar an entrepreneur. For example, the course in Seed Technology should be so restructured that it becomes, "Seed Technology and Business". This will make it unnecessary for the scholar to go to a Business School after earning an agricultural degree.

Some years ago, the Government of India launched a programme for enabling farm graduates to start agri-clinics and agri-business centres. This programme is yet to attract the interest of educated youth to the degree originally expected. It is hence time that the programme is restructured based on the lessons learnt. Ideally a group of four to five farm graduates, who have specialised in agriculture, animal husbandry, fisheries, agri-business and home science could jointly launch an agri-clinic-cum-agri-business centre in every block in the country. Agri-clinics will provide the services needed during the production phase of farming, while the agri-business centre will cater to the needs of farm families during the post-harvest phase of agriculture. Thus, farm women and men can be assisted during the entire cropping cycle, starting with sowing and extending up to value addition and marketing. The multi-disciplinary expertise available within the group of young entrepreneurs will help them to serve farm families in a holistic manner. The Home Science graduate can pay particular attention to nutrition and food safety and processing and help a group of farm women to start a Food Processing Park. The group should also assist farm families to achieve economy and power of scale both during the production and post-harvest phases of farming. Such an integrated centre can be called the Agricultural Transformation Centre.

There are several opportunities for such young entrepreuners to initiate programmes in the fields of soil health enhancement, plant and animal health care, seed technology and hybrid seed production. Climate resilient agriculture is another area needing attention. In dry-farming areas, methods of rainwater harvesting and storage and watershed management as well as the improvement of soil physics, chemistry and microbiology, need to be spread widely. The cultivation of fertiliser trees, which can enrich soil fertility and help to improve soil carbon sequestration and storage, can be promoted under the Green India Mission as well as the Mahatma Gandhi National Rural Employment Guarantee programme. A few fertiliser trees, a jal kund (water harvesting pond) and a biogas plant in every farm will enormously help improve the productivity and profitability of dry-land farming. In addition, they will contribute to climate change mitigation.

The Yuva Kisans or young farmers can also help Women Self-help Groups to manufacture and sell the biological software essential for sustainable agriculture. These will include biofertilisers, biopesticides and vermiculture. The Fisheries graduate can promote both inland and marine aquaculture, using low external input sustainable aquaculture (LEISA) techniques. Feed and seed are the important requirements for successful aquaculture and trained youth can promote their production at the local level. They can train rural families in induced breeding of fish and spread quality and food safety literacy.

Similar opportunities exist in the fields of animal husbandry. Improved technologies of small-scale poultry and dairy farming can be introduced.

Codex Alimentarius standards of food safety can be popularised in the case of perishable commodities. For this purpose, the young farmers should establish Gyan Chaupals or Village Knowledge Centres. Such Centres will be based on the integrated use of the internet, FM radio and mobile telephony. For example, artisanal fishermen going out into the sea in small boats can now be empowered with information on wave heights at different distances from the shoreline and also on the location of fish shoals. Such techniques will help to transform the lives of small-scale fisher families.

In the services sector designed to meet the demand-driven needs of farming families, an important one is soil and water quality testing. Young farmers can organise mobile soil-cum-water quality testing work and go from village to village in the area of their operation and issue a Farm Health Passbook to every family. The Farm Health Passbook will contain information on soil health, water quality, and crop and animal diseases, so that the farm family has access to integrated information on all aspects of farm health. Very effective and reliable soil-testing kits are now available. This will help rural families to utilise in an effective manner the nutrient-based subsidy introduced by government. Similarly, young educated youth could help rural communities to organise Gene, Seed, Grain, Water Banks, thereby linking conservation, cultivation, consumption and commerce in a mutually reinforcing manner. It is only through the provision of such services that we can achieve the goal of improving the economic well-being of rural families. Young farmers can also operate Climate Risk Management Centres, which will help farmers to maximise the benefits of a good monsoon, and minimise the adverse impact of unfavourable weather.

Educated youth can help to introduce in rural India the benefits of information, space, nuclear, bio- and eco-technologies. Eco-technology involves the blend of traditional wisdom and frontier technology. This is the pathway to sustainable agriculture and food security, as well as agrarian prosperity. If educated youth choose to live in villages and launch the new agriculture movement, based on the integrated application of science and social wisdom, our untapped demographic dividend will become our greatest strength.

❑❑❑

Chapter 7

Legislation for Food Security

The National Food Security Bill 2011, which is now on the website of the Union Ministry of Food and Consumer Affairs for public comments, aims to make the Right to Food a legal right. During the last seven years, the country has witnessed a transition from political patronage to legal entitlements in the case of education, information, work and land rights in tribal areas. When finally enacted, the Food Security Bill will be the brightest jewel in the crown of Indian democracy. Therefore, public scrutiny of the draft Bill is important. The draft Bill mentions that its aim is "to provide for food and nutritional security in human life-cycle approach by ensuring access to adequate quantity of quality food at affordable prices, for people to live a life with dignity". Unfortunately the Bill is its present form will not be able to fulfill this inspiring objective.

The legal commitment contained in the Bill implies that every child, woman and man should have physical and economic access to balanced diet on the basis of a life-cycle approach, i.e., from conception to cremation. Nutrition security involves access not only to the needed calories and protein, but also to micro-nutrients like iron, iodine, zinc, vitamin A, and vitamin B_{12}. In addition, clean drinking water, sanitation and primary health care are essential for ensuring that food is properly assimilated in the body. Food and nutrition security will thus need concurrent attention to both food and non-food factors. Obviously, every requirement cannot become overnight a legal right. Therefore, government has confined the legal right only to economic access with reference to certain quantities of grain, like rice, wheat, and nutri-cereals such as *ragi, bajra, jowar*, maize, etc. The Bill provides for common and differentiated rights. The common rights are designed to ensure that every citizen in the country has access to food which is the first among the hierarchical needs of human beings. The differentiated rights relate to quantity and cost of the food to be provided to the general category of citizens who are not in need of the same kind of social support as those listed under the priority category. The draft Bill circulated for comments requires considerable improvement which undoubtedly will be made by the Standing Committee of Parliament which will examine the Bill before it goes for adoption.

In my view, the Bill to achieve its purpose of nutrition security for all coupled with respect for human dignity will need the following structure.

- **Legal Entitlements:** This will begin with pregnant mothers in order to avoid maternal and foetal undernutrition. The present Integrated Child Development Services (ICDS) could be divided into two segments from the point of view of the age of the child. The first thousand days starting from conception are exceedingly important for brain development in the child and for avoiding low birth weight at the time of delivery. This is the neglected part of the present ICDS and this is why we have nearly every fourth child born in the country having a birth weight below 2.5 kg. Such low birth weight children have many handicaps in later life, including impaired cognitive abilities. The older children can be provided nutritious noon meals and also other forms of nutrition support like milk, and nutri-biscuits. As far as adults are concerned, the Bill provides for the provision of seven kilograms of food grains per person per month in the case of priority households. The price will not exceed Rs.3, 2 or 1 per kg for rice, wheat and nutri-cereals, respectively. The draft Bill also makes provision for providing support to special groups such as destitutes and homeless persons. The general households not requiring the same kind of social support as priority households will be provided with 3 kg of foodgrains per person per month at a price not exceeding 50 per cent of the minimum support price.

- **Enabling Provisions:** Food security has three dimensions, namely, **availability** of food, which is a function of production, **access** to food which is a function of purchasing power, and **absorption** of food in the body, which is a function of the availability of clean drinking water, sanitation and primary health care. Therefore the Bill, to achieve the goal of food and nutrition security, should emphasise the need for effective implementation and close monitoring of the following schemes:

  a. Ensuring adequate availability of food by implementing the provisions of the National Policy for Farmers placed in Parliament in November 2007, as well as of the schemes designed to stimulate higher production such as the Rashtriya Krishi Vikas Yojana, National Food Security Mission, National Horticulture Mission and Mahila Kisan Sashaktikaran Pariyojana.

  b. Effective implementation of the Rajiv Gandhi Drinking Water Mission, Total Sanitation Programme and Rural Health Mission

  c. Mainstreaming nutrition in the horticulture mission in order to provide horticultural remedies to nutritional maladies, such as deficiency of iron, iodine, vitamin A, etc. For example, a combination of *Moringa* (drumstick) and *ragi* or *bajra* can provide all the needed macro- and micro-nutrients.

  The delivery of these provisions must be made in a "deliver as one mode" in order to ensure synergy among the different components of food security.

- **Reform of the Public Distribution System:** Several successful models are already available, as for example in Chattisgarh, Tamil Nadu and Kerala. Modern technology like Smart Cards could be used to prevent leakages in delivery. In the ultimate analysis, a corruption-free India will be an essential prerequisite for a hunger-free India. More effective use of Gram Sabhas and elected local bodies will make a useful contribution to checking corruption both in the identification of priority groups and in stopping pilferage.

- **Building the necessary infrastructure**: A Food Security Bill can be implemented only with the help of home-grown food. In other words, the well-being of farmers will be essential for ensuring food security. It will be tempting to suggest the distribution of cash instead of grain for overcoming problems arising from inadequate growth in food production. Distribution of cash rather than grain will lead to a loss of interest in public procurement. If procurement goes down, production will also go down. This will be disastrous in a country where nearly two-thirds of the population depend on agriculture for their livelihood. Enhancing small farm productivity is the most effective method of ending endemic hunger in rural India. Hence we should start building proper storage structures. I have often suggested that we should build at least 50 ultra-modern grids of silos, each grid capable of storing safely one million tonnes of foodgrains at 50 different locations. In other words, government will always have about 50 million tonnes of foodgrains stored in different parts of the country, particularly in the North East and other remote areas, which can ensure uninterrupted supply of foodgrains to the public in all parts of the country.

The present draft is a good beginning to seriously address issues relating to poverty- induced chronic hunger. We should however make a bold and imaginative attempt to rid the country of chronic hunger and malnutrition. If we do this, we will meet with a win-win situation both for agricultural progress and freedom from hunger.

❑❑❑

Chapter 8

How to Resolve the Crisis of Indian Agriculture

The post-independence history of our agriculture can be broadly grouped into four periods. Before describing them, I should mention that during the colonial era famines were frequent and Famine Commissions were abundant. The growth rate in food production from 1900 to 1947 was hardly 0.1 per cent. Most of the important institutional developments in agriculture emanated from the recommendations of Famine Commissions. The great Bengal Famine of 1942-43 provided the backdrop to India's independence. It is to the credit of Independent India that famines of this kind have not been allowed to occur, although our population has grown from 350 million in 1947 to over a billion now.

Phase I (1947-64)

This was the Jawaharlal Nehru era where the major emphasis was on the development of infrastructure for scientific agriculture. The steps taken included the establishment of fertiliser and pesticide factories, construction of large multi-purpose irrigation-cum-power projects, organisation of community development and national extension programmes and, above all, the setting up of agricultural universities, beginning with the Pant Nagar

University established in 1958 as well as new agricultural research institutions as, for example, the Central Rice Research Institute at Cuttack and the Central Potato Research Institute, Shimla.

During this period, the population of the country started increasing by over 3 per cent per year as a result of both the steps taken to strengthen public health care systems and advances in preventive and curative medicine. The growth in food production was inadequate to meet the consumption needs of the growing population, and food imports became essential. Such food imports, largely under the PL-480 programme of the United States, touched a peak of 10 million tonnes in 1966.

Phase II (1965-1985)

This period coincides with the leadership of Lal Bahadur Shastri and Indira Gandhi with Morarji Desai and Charan Singh serving as Prime Ministers during 1977-79. The emphasis was on maximising the benefits of the infrastructure created during Phase I, particularly in the areas of irrigation and technology transfer. Major gaps in the strategies adopted during Phase I were filled, as for example the introduction of semi-dwarf high-yielding varieties of wheat and rice, which could utilise sunlight, water and nutrients more efficiently and yield 2 to 3 times more than the strains included in the Intensive Agriculture District Programme (IADP) of the early 1960s. This period also saw the re-organisation and strengthening of agricultural research, education and extension and the creation of institutions for providing farmers assured marketing opportunities and remunerative prices for their produce. A National Bank for Agriculture and Rural Development (NABARD) was set up. All these steps led to a quantum jump in the productivity and production of crops like wheat and rice, a phenomenon christened in 1968 as the Green Revolution C. C. Subramanian (1964-67) and later Jagjivan Ram provided the necessary public policy guidance and support.

The Green Revolution generated a mood of self-confidence in our agricultural capability. The gains were consolidated during the Sixth Five Year Plan period (1980-85) when, for the first time, agricultural growth rate exceeded the general economic growth rate. Also, the growth rate in food production exceeded that of population. The Sixth Plan achievement illustrates the benefits arising from farmer-centred priorities in investment and in the overall agricultural production strategy.

Phase III (1985-2000)

This was the era of Rajiv Gandhi, P. V. Narasimha Rao and Atal Bihari Vajpayee, with several other Prime Ministers serving for short periods.

This phase was characterised by greater emphasis on the production of pulses and oilseeds as well as of vegetables, fruits and milk. Rajiv Gandhi introduced organisational innovations like Technology Missions which resulted in a rapid rise in oilseed production. The Mission approach involves concurrent attention to conservation, cultivation, consumption and commerce. Rain-fed areas and wastelands received greater attention and a Wasteland Development Board was set up. Wherever an end-to-end approach was introduced, involving attention to all links in the production-consumption chain, progress was steady and sometimes striking, as in the case of milk and egg production. This period ended with large grain reserves with government, with the media highlighting the co-existence of "grain mountains and hungry millions". This period also saw a gradual decline in public investment in irrigation and the infrastructure essential for agricultural progress as well as a gradual collapse of the cooperative credit system.

Phase IV (2001 to the present day)

Despite the efforts of Prime Ministers Atal Bihari Vajpayee and Manmohan Singh, this phase is best described as one characterised by policy fatigue resulting in technology, extension and production fatigues. No wonder that farmers who keep others alive, are now forced to take their own lives and 40 per cent of them want to quit farming, if there is an alternative option. The agricultural decline is taking place at a time when international prices of major foodgrains are going up steeply, partly due to the use of grains for ethanol production. Land for food versus fuel is becoming a major issue. For example, the export price of wheat rose from US$197 per tonne to US$263 per tonne in two years. Maize price has gone up from about US$100 per tonne in 2005 to US$166 per tonne. International trade is also becoming free, but not fair. Compounding these problems is the possibility of adverse changes in rainfall, temperature and sea level as a result of global warming. The melting of Himalayan ice and glaciers will result in floods of unprecedented dimensions in north India. If the rate of agricultural production does not remain above population growth and if the public distribution system is starved of grains, there is every likelihood of going back to the pre-independence scenario of recurrent famines. The grain mountains have disappeared and we are today in the era of diminishing grain reserves, escalating prices and persistence of widespread undernutrition.

Where do we go from here ?

The Green Revolution of the 1960s was the result of synergy among technology, public policy and farmers' enthusiasm. The post 60th anniversary era in agriculture will depend upon our determination to implement Jawaharlal Nehru's exhortation, "Everything else can wait, but not agriculture" in both letter and spirit.

If farm ecology and economics go wrong, nothing else will go right in agriculture. This is the principal message of the current agrarian crisis, which is very likely to spread if the economics of small-scale farming is not improved. At the same time, State governments should not promote policies for ecocides (ecological suicides) such as free electricity for pumping groundwater leading to the exhaustion of the aquifer.

How can we resolve the crisis ? The first and foremost priority should go to making the era of farmers' suicide history.

About 35 districts identified by the Union Ministry of Agriculture as most affected by the agrarian crisis should be developed into Special Agricultural Zones (SAZ), where integrated attention will be paid to natural resources conservation and enhancement, eco-farming, improved local-level consumption to overcome malnutrition and pro-small farmer commerce. Most of these areas are rain-fed and attention will have to be paid to the generation of multiple livelihood opportunities. These areas require the joint efforts of agricultural scientists, extension agencies, policy makers and mass media. Unless the various government departments/ministries dealing with agriculture, animal husbandry, fisheries, forestry, environment, agro-processing and agri-business, irrigation, commerce, rural development and finance, work on the principles of convergence and synergy, it will be difficult to find lasting solutions to the problems of small farmers. The major purpose of a Special Agricultural Zone is ecological restoration and the strengthening of the work and income security of farm families with about 1 hectare or less of land. While the Special Economic Zone (SEZ) is designed to enhance trade and export income involving mega-investment by the private sector industry, SAZ is needed for saving the lives and livelihoods of small farmers and landless labour by providing key centralised services to support decentralised small-scale production as well as market and income security. The SAZ concept will provide an effective method of ending farmers' suicides by creating a platform for collective action by all the concerned departments and agencies of the Central and State governments, private sector industry and civil society organisations. The present relief measures are fragmented both in design and implementation and unless they are replaced with a holistic approach, with special emphasis on minimising risks and maximising net income, the crisis will get worse.

While carefully designed SAZs can help to end the era of farmers' suicides, the emerging larger agricultural production and food security crisis can be managed if the following steps are taken to achieve an ever-green revolution in agriculture leading to the enhancement of productivity in perpetuity without associated ecological harm. The five basic components of an ever-green revolution strategy are the following:

- Conservation of prime farm land for agriculture and soil health care and enhancement, and the issue of Soil Health Cards indicating the organic matter and macro- and micro-nutrient status of the soil
- Water harvesting, management and conjunctive use of surface, rain, ground and treated effluent water and safeguarding water quality
- Credit and insurance reform
- Low-risk and environmentally-friendly green technologies (such as integrated pest and nutrient management) and the provision of the needed inputs at the right time and place and at affordable cost
- Assured and remunerative marketing

These five steps need to be taken and implemented in an integrated manner, so that we generate an Ever-green Revolution Symphony.

While such course of action is recommended in all farming zones, differentiated steps are however needed in the following three areas:

First, we must defend the gains already made in the Green Revolution areas of the Punjab, Haryana and western Uttar Pradesh. This heartland of the Green Revolution or India's fertile crescent is in a state of acute ecological and economic distress. Conservation farming and green agriculture should replace exploitative agriculture. Public policies promoting ecocides should be withdrawn and replaced with incentives for conservation farming. This region will continue to remain a major source of foodgrains for the public distribution system and hence needs urgent attention.

Second, we must extend the gains to additional areas like Bihar and the entire eastern part of India which possess good soil and water resources, as well as to rain-fed, hill and coastal areas. A second fertile crescent can be created immediately in the Bihar, eastern Uttar Pradesh, Chattisgarh, West Bengal and Assam regions, where the untapped production reservoir even with technologies on the shelf is high.

Finally, we should make new gains, particularly in the areas of diversification of farming systems and value addition. There is now a mismatch between production and post-harvest technologies. This should be ended. A quality literacy and value-addition movement should be launched.

The National Commission on Farmers has outlined a detailed strategy for achieving the above goals. A draft National Policy for Farmers has also been provided by NCF, which, if adopted, will help to make the growth rate in the net income of farmers as the principal criterion for measuring agricultural progress.

Farmers are ready to help the nation. Are we ready to help them ?

❑❑❑

Chapter 9

Gandhiji's Plea for a Hunger-Free India: Current Reality and Way to Progress

'To a people famishing and idle, the only acceptable form in which God can dare appear is work and promise of food as wages' — these were the words of Mahatma Gandhi when he was healing the wounds arising from the Hindu-Muslim divide at Naokhali in 1946. He thus stressed the symbiotic bonds among work, income and food security.

Eradication of hunger and poverty is also the first among the UN Millennium Development Goals, which in my view represent a Global Common Minimum Programme for human security and well-being.

Achieving this goal is fundamental for achieving the other goals relating to education, gender equality, child mortality, control of HIV/ AIDS, malaria and other diseases and, above all, environmental sustainability. Unfortunately, progress in the reduction of hunger and poverty is poor in most developing countries excepting a few like China. FAO estimates that as a result of the rise in the prices of basic staples, about 75 million more people have been added to the list of those going to bed hungry during 2007. As a single nation, we in India have the largest number of malnourished persons in the world.

Fortunately, all Indian political parties are committed to the eradication of hunger and achieving the UN Millennium Development Goals in the area of hunger and poverty elimination. Our former Prime Minster Atal Bihari Vajpayee, for example, said in 2001 on the occasion of the release of the

Food Insecurity Atlas of Rural India[1] prepared by MSSRF and the World Food Programme (WFP): 'The sacred mission of a hunger-free India needs the cooperative efforts of the Central and State governments, non-governmental organisations, international agencies and all our citizens. We can indeed banish hunger from our country in a short time.' Prime Minister Manmohan Singh has reiterated this resolve: 'The problem of malnutrition is a curse that we must remove. Our efforts to provide every child with access to education, and to giving equal status to women and to improve health care services for all citizens will continue.' How can we convert this political resolve into practical accomplishment ?

Nearly 70 per cent of India's population lives in villages, where the main source of livelihood is agriculture, comprising crop and animal husbandry, fisheries, agro-forestry and agro-processing. Enhancing the productivity of small farms and thereby the marketable surplus available for earning cash income is a powerful method of reducing malnutrition among over 500 million members of small farm families, who fall under the category of producer-consumers. Accelerated agricultural progress helps to strengthen both national food security and household nutrition security.

In 2007, on the occasion of the 60th anniversary of our Independence, a broad-based Coalition for Sustainable Nutrition Security in India was formed at a meeting held at MSSRF. The Coalition, comprising national, USAID (United States Agency for International Development) and UN organizations, has prepared a report titled "Overcoming the Curse of Malnutrition: A Leadership Agenda for Action". The following 5-point action plan has been the outcome of the discussions.

1. Institutional structures for public policy and coordinated action in nutrition

Overcoming malnutrition requires concurrent attention to food (macro- and micronutrients, clean drinking water) and non-food factors (sanitation, environmental hygiene, primary health care, nutrition literacy and work and income security). Nutrition security at the level of each individual is vital for providing an opportunity for a healthy and productive life. Achieving the goal of nutrition security for all will need the fusion of political will and action, professional skill and people's participation. Such a coalition of policy makers,

[1] MSSRF-WFP (2001). *Food Insecurity Atlas of Rural India.* Chennai: M. S. Swaminathan Research Foundation.

professionals and citizens will have to start from the village and go up to the national level. The following consultative, policy oversight and monitoring structures are suggested.

- Panchayat/ Nagarpalika/ Local Body

Council for Freedom from Hunger, established by Gram Sabhas/ Local Bodies, with one man and one woman from each village being trained as hunger fighters.

- State / Union Territory

State Level Committee on Nutrition Security, chaired by the Chief Minister, with all concerned Ministers and representatives of civil society organisations, corporate sector and mass media

- National Level

Cabinet Committee for Nutrition Security, chaired by the Prime Minister

A system for horizontal linkages among these three levels of action will have to be developed.

2. Learning for success: converting the unique into the universal

Nothing succeeds like success. Therefore it is important to learn from successful examples of the elimination of malnutrition. Thailand, Vietnam, Brazil and China have achieved significant reductions in the level of malnutrition through an integrated strategy involving education, social mobilisation and nutrition safety nets. Thailand brought about a substantial reduction in infant mortality rate (IMR) speedily through a large cadre of community health workers. At the national level, Kerala and Tamil Nadu have been successful in reducing malnutrition. A unique combination of Integrated Child Development Services (ICDS) and the Tamil Nadu Integrated Nutrition Project (TINP) was launched in Tamil Nadu where TINP identified a community worker to concentrate on families with children between 0-3 years of age. Special attention to pregnant women belonging to economically underprivileged families is also essential for avoiding the occurrence of babies with low birthweight (LBW).

In Tamil Nadu, the cooperative sector runs 96 per cent of ration shops and the remaining are managed by panchayats and Womens' Self-Help Groups. One big advantage of using the cooperative system is that credit facility is available to purchase grain from PDS. From 1982, Tamil Nadu has been operating a universal noon-meal programme for school children, which now covers old age pensioners, destitutes, widows and pregnant women. Various indicators of malnutrition show a downward trend in Tamil Nadu. For example, the incidence of severe malnutrition (Grades III and IV) among children aged 0-36 months declined for 12.3 per cent in 1983 to 0.3 per cent in 2000. In Kerala, there has been effective monitoring of quality of supply, timeliness and other features of PDS by panchayats and social activists. It would be useful to replicate such effective measures to combat malnutrition in all the States.

Successful programming experience and health and nutrition evidence show that breaking the curse of malnutrition will require focusing on two important target groups: children under two years of age and women, especially adolescent girls, pregnant women and lactating women. Rates of child malnutrition in India are among the highest in the world and, more worrisome, the nutrition situation of Indian children has not improved significantly over the last decade. In 1998-99, the prevalence of child underweight was 43 per cent (NFHS-2); in 2005-06 the prevalence of child underweight was 40 per cent (NFHS-3); this is a mere 0.5 percentage point annual decrease over the six years. Population is increasing by over 16 million every year and hence the number of malnourished children is actually increasing. This is a matter for serious national concern. Although preventing malnutrition needs to be the focus of our policy and programme action, we have many children currently suffering from severe acute malnutrition. For these children, adequate treatment must be made available as a matter of entitlement.

The first two years of life represent the critical window of opportunity to break the inter-generational cycle of malnutrition. If this critical window of opportunity is missed, child malnutrition will continue to self-perpetuate and malnourished girl children will continue to grow to become malnourished women who give birth to low birth-weight infants who are poorly fed in the first two years of life. This vicious inter-generational cycle of malnutrition requires a concerted investment effort in improving the nutrition situation of infants and young children from conception through the first two years of life. Investing in girls and women has also shown the potential for being transformational for the health, nutrition and well-being of the entire household and community.

State governments should develop a 'Hunger Free State' strategy, which adopts a lifecycle approach to the delivery of nutrition support to reach the key target groups and vulnerable sections of the population. In such a strategy, the strategies of TINP and ICDS could be suitably integrated, with a special programme to prevent maternal, foetal and young child malnutrition. ICDS should be redesigned into time frames: the fist 1000 days of a child's life from conception to the first 2 years and the next 1000 days.

3. Action at local level: community food- and nutrition-security systems

Community food- and nutrition-security systems including the setting up of Grain, Seed, Fodder and Water Banks, can be promoted by local bodies. The food basket should be widened, so as to include a wide range of millets like ragi, legumes, vegetables and tubers. The Panchayat/Local Council for Freedom from Hunger could mobilise the needed technological and credit support for establishing the Grain, Seed Fodder and Water Banks. Wherever hidden hunger from the deficiency of iron, folic acid, iodine, zinc and vitamin A in the diet is endemic, food–cum–micronutrient supplementation and appropriate and effective fortification approaches (as for example, iodine- and iron-fortified salt) can be adopted. Every Panchayat/Local Council for Freedom from Hunger could invite a Home Science Graduate in the area to serve as a Nutrition Advisor.

4. Action at State level: coordinating nutrition-security initiatives

The State Level Committee on Nutrition Security chaired by the Chief Minister of the State should facilitate the implementation of the numerous ongoing nutrition safety net programmes (national, bilateral and international) in a coordinated and mutually reinforcing manner, in order to generate synergy and thereby maximise the benefits from the available resources. The Horticulture Mission provides a unique opportunity for providing local-level horticultural remedies to major nutritional maladies. Overcoming micronutrient malnutrition and intestinal load of infection are urgent tasks. State governments should launch a nutrition literacy movement and set up media coalitions for nutrition security for improved nutrition awareness. Such a media coalition should include representatives of print media, audio and video channels, new media including the internet, and traditional media like folk dance, music, and street plays.

5. Action at national level: mainstream nutrition in National Missions

At the national level, the most urgent tasks relate to including nutrition outcome indicators and targets in all major Missions in the field of agriculture and rural development. Large national programmes like the Rashtriya Krishi Vikas Yojana (Rs 25,000 crore), the National Horticulture Mission (Rs 20,000 crore) and the National Food Security Mission (Rs 5,000 crore) should have a Nutrition Advisory Board, so that cropping and farming systems are anchored on the principle of food-based nutrition security. Also, the delivery of various nutrition safety net programmes should be organised on a lifecycle basis so that integrated attention can be given to the needs of an individual from birth to death. Similarly, the NREGA (National Rural Employment Guarantee Act) sites, where mostly illiterate women and men work on unskilled jobs, should have a nutrition clinic operated by a knowledgeable person and a PDS facility. If food is not available at affordable prices at NREGA sites, most of the money earned will go to purchasing staple foods at high cost and undernutrition will persist. Gyan Chaupals can run adult nutrition literacy programmes based on computer-aided learning technology.

The National Rural Health Mission supported by a large number of ASHAs (Accredited Social Health Activists) offers an uncommon opportunity for strengthening health and nutrition security. It is worthwhile to consider methods of adding a nutrition component to this Mission and thereby launching an Integrated National Rural Health and Nutrition Mission. Obviously such an integrated mission is equally important for urban areas, since urban food and nutrition insecurity is equally great, as revealed by the *Food Insecurity Atlas of Urban India* prepared by MSSRF and WFP.[2]

---

[2] MSSRF-WFP (2002). *Food Insecurity Atlas of Urban India*. Chennai: M. S. Swaminathan Research Foundation.

In an Integrated National Rural / Urban Health and Nutrition Mission, civil society organisations and the corporate sector can play a significant role. The corporate sector under their Corporate Social Responsibility programmes can help to achieve "Health and Nutrition for All" in the areas where they operate. It is also clear that diseases like HIV/AIDS and tuberculosis can be cured only if there is a nutrition-cum-drug- based approach. Nutrition should hence be integrated into the national disease control programmes.

Integrating nutrition in relevant National Missions will accelerate the pace of progress in providing an opportunity for every child, woman and man in the country for a healthy and productive life. Without such an accomplishment a sustainable foundation for inclusive economic growth cannot be laid. This is the number one challenge facing the nation.

To conclude, we have the political will, technical skill and resources to achieve Gandhiji's goal of a hunger-free India. Can we now bring all the needed inputs and action together so that nutrition for all becomes a reality? We should walk our talk and not further postpone erasing the stigma and shame associated with our country being the home for the largest number of malnourished in the world.

❑❑❑

Chapter 10

Wheat Imports and Food Security

The pros and cons of wheat imports are now a matter of public, political and media concern and debate. I have been quoted as both favouring imports and opposing it. Since this issue is a complex one with short- and long-term implications, I would like to explain my personal position.

The wheat import plan announced by the Government of India is in response to the need for maintaining adequate food stocks both for the purpose of food security and for feeding the public distribution system (PDS). Following the Wheat Revolution in 1968, Indira Gandhi decided to build a sufficient buffer stock of foodgrains under government control in order to ensure that the basic staples are available at reasonable prices even under conditions of unfavourable monsoon. She was also deeply conscious of the linkage between food security and sovereignty in foreign policy. Maintaining adequate food reserves is an absolute must from the point of view of avoiding both panic purchase and famines, particularly at times when there are indications of aberrant monsoon behaviour. Some years ago the Government of India had over 60 million tonnes of foodgrain reserves and substantial quantities were exported to reduce the cost of maintaining stocks of that order. There was criticism at that time that while we were exporting foodgrains, millions of children, women and men were going to bed partially hungry in our country. The situation was described in the media as "grain mountains and hungry millions". Even now we have the largest number of malnourished children in the world. Only the grain mountains have disappeared.

The Minister for Agriculture and Food has explained that imports have become essential to build a buffer stock. Normally much of the surplus grain is bought by government agencies like the Food Corporation of India at the support price announced by government. Now private parties including large corporations have also entered the grain market and have been able to procure a part of the surplus grains at higher price. The farmers are certainly happy when they are able to get prices higher than the price offered by the government. However, this situation has led to a shortfall in procurement target, thus necessitating wheat imports.

Maintaining food security for our population of over a billion is a fundamental obligation of government. Fortunately, we have sufficient foreign exchange reserves and hence imports can be resorted to where there is no other alternative to replenish the buffer stock and operate PDS. What is important is to draw appropriate lessons from this experience and undertake the development of a long-term policy for home-grown-food-based food security, where both the public and private sectors perform well-defined and mutually-reinforcing roles. Such a strategy should be designed to promote the availability of the staple grains and food commodities needed for nutrition security at the right time and place as well as at an affordable price. The private sector should develop its own code of conduct and should not give the impression that for short-term financial gains, the health and nutritional security of millions of children, women and men are being sacrificed. There is need for a transparent and well-defined code of conduct in the areas related to the purchase and trade of basic foodgrains. The nation as a whole must take pride in an effective food security system.

There have also been adverse comments on the removal of the 10 per cent tariff imposed on the import of pulses. In our country, the situation with reference to producers and consumers is different from that in the industrialised nations. Nearly 65 per cent of the consumers in India are also producers, many of them operating farms of 1 ha and less in size. Rural malnutrition is more widespread than urban malnutrition. Small farm families depend for their livelihood on income from the sale of whatever surplus quantities of foodgrains, vegetables, fruits and milk they may have for the market. Pulses and oilseeds are predominant crops of rain-fed and dry farming areas. Productivity is low since efforts in taking new technologies and seeds are poor or inadequate, although there is a huge stockpile of underutilised technologies including new varieties. If we continue the practice of importing pulses and oilseeds, dry farming areas will continue to languish in poverty and malnutrition. The linkage between low small farm productivity and the persistence of poverty and malnutrition is very strong. Therefore, the sooner we revise our import policies in relation to pulses and oilseeds and divert our attention to helping the millions of farmers toiling in rain-fed areas, the greater will be the possibility of substantially reducing hunger and poverty in the country. Whenever there is a good crop of pulses or oilseeds like the one in mustard, farmers suffer due to lack of assured and remunerative marketing opportunities. The interests of the producer-consumers need greater protection than the interests of trader-importers.

The achievement of environmentally, economically and socially sustainable food security or "food for all and for ever" must be at the top of a national common minimum programme. The strategy for achieving this goal must be the result of a political consensus and should transcend political frontiers. It would therefore be useful to constitute a National Food Security Board chaired by the Prime Minister. In addition to the Food and Agriculture Minister, such a Board could include other concerned Union Ministers, a few Chief Ministers of food-surplus and food-deficit States, leaders of political parties, food and gender specialists and representatives of farmers' associations and the print, electronic and new media. Such a Board can help to develop and implement an action plan for providing economic, physical and social access to a balanced diet and safe drinking water to every child, woman and man in the country. The National Commission on Farmers in its second Report submitted in the middle of 2005 has offered suggestions for bringing about a paradigm shift from food security at the national level to nutrition security at the level of each individual.

To sum up, imports or exports of foodgrains may be necessary from time to time, but the bottom line of our import-export policies must be the livelihood security of both the farm and non-farm populations of rural India who constitute 70 per cent of our population. We are confronted with the need to safeguard the food security requirements of both resource-poor farmers and resource-poor consumers. The bulk of such resource-poor farmers are small or marginal farmers and landless agricultural labour in unirrigated areas. It is these linkages which need to be understood and attended to. The proposed National Food Security Board can attend to these complex linkages in a holistic manner and develop and implement a transparent national food security policy.

❑❑❑

Chapter 11

Distribute, Procure, Store and Sow — Steps to Food Security

The goal of food for all can be achieved only through sustained efforts in producing, saving and sharing food grains.

The Supreme Court of India has rendered great service by arousing public, professional and political concern about the co-existence of rotting grain mountains and mounting hungry mouths. In several African countries hunger is increasing because food is either not available in the market, or is too expensive for the poor. Food inflation is showing no signs of abating. In our country, chronic hunger is largely poverty induced. The progress made in achieving the UN Millennium Development Goal No. 1, namely, reducing hunger and poverty by half by the year 2015, has been far from satisfactory and the available data indicate that we may have years and years to go before we achieve this target. Globally, the number of persons going to bed hungry has increased from 800 million in the year 2000 to over one billion. The position is likely to get worse in the near term, since the prices of wheat, rice and maize are going up in the global market. Adverse growing conditions in Russia, Canada and Australia are partly responsible for the recent escalation in grain prices. Nearer home, Pakistan is recovering from serious floods. According to a recent UN report, 3.2 million ha of standing crops and 200,000 heads of livestock have been lost. Pakistan may need large quantities of wheat seed for *rabi* sowing for which we are the only suitable source. In our country, Jharkhand, Bihar and West Bengal had until recently experienced deficit rainfall. Often, early drought is accompanied by severe flood during September-October and hence we need both drought and flood codes to be put into operation at different times during the southwest monsoon period, particularly in Bihar and Assam. Local level Seed and Grain Banks should be built to ensure crop and food security under conditions of unpredictable monsoon.

Among the steps needed to concurrently address the alleviation of hunger and safeguarding farmers' incomes, the following four need urgent attention :

Distribute the grains for which there is no safe storage facility

Gandhiji emphasised that hunger should be overcome without eroding human dignity. He wanted every Indian to have an opportunity to earn his or her daily bread. There are however seriously disadvantaged sections of our population like orphans, street children, widows, old and infirm persons, pregnant women suffering from anaemia, children in the age group 0 to 2 belonging to poor families, and those affected by leprosy, tuberculosis and HIV/AIDS, who need to be provided with food free of cost. In the case of diseases like leprosy, tubercuosis and HIV/AIDs, we need a food-cum-drug approach since many of the patients suffering from such ailments tend to be undernourished, thereby reducing the beneficial impact of the treatment. A National Committee I chaired had made this recommendation nearly 30 years ago in the case of leprosy. The recent decision of the Government of India to provide 25 lakhs of additional foodgrains for BPL families is a welcome step. This should be supplemented by providing free food to those suffering from extreme destitution and poverty through delivery systems like community kitchens run by agencies not likely to be affected by corruption.

Food stocks exposed to rain and consequently having high moisture content are likely to get infected with *Aspergillus sp.*, leading to the development of mycotoxins. Hence, they should not be distributed among the poor, without prior testing by institutions like the National Institute of Nutrition, Hyderabad and the Central Food Technological Research Institute, Mysore. We should not add to the nutritional problems of the poor by offering them grains containing aflatoxins. Food losses due to poor storage should be measured both in quantitative and qualitative terms. Subject to such screening, foodgrains fit for human consumption are best distributed free among the most deprived sections of our society throughout the country. To begin with, about 5 million tonnes of wheat and rice could be allotted for this purpose from the stocks for which good storage conditions are not available.

Procure *kharif* crops

The recommendation of the National Commission on Farmers (NCF) that the minimum support price (MSP) should be C-2 (i.e., total cost of production) plus 50 per cent is yet to be implemented. Government has however been moving in the direction of a MSP which is relatively fair to the farmers. The MSP announced for rice and pulses is reasonably attractive and consequently, the production of pulses, rice, *jowar, bajra,* maize and oilseeds is likely to be good. Over 20 million tonnes of rice will have to be procured soon. Hence, no further time should be lost in making arrangements for the safe storage of the purchased grains. Also, the procurement by Central and State agencies should extend to pulses and crops like *jowar, bajra, ragi* and maize, so that there is a diversification of the food basket. Procurement at remunerative price is the key to keeping up farmers' interest in farming. Unlike the Right to Information Act which can be implemented with the help of files, the right to food can be implemented only with the help of farmers.

The gap between potential and actual yields is high in pulses, oilseeds and the other crops sown in rain-fed areas. Instead of going to Canada and Australia for procuring pulses for us, the State and Central governments should acquire them directly from our farmers, as is being done in the case of wheat and rice. The 60,000 Pulses and Oilseed Villages in rain-fed areas, for which provision of funds has been made in the Union Budget for 2010-11, should be designed on a systems approach with concurrent attention to all the links in the production, protection, procurement and consumption chain, as envisaged under the Rajiv Gandhi Pulses and Oilseed Missions of the 1980s.

Store safely

From the Vedic period, food has been invested with an aura of respect and reverence. It is sad that this respect has been destroyed by those in charge of procurement and storage, particularly when we are classified as a nation with the largest number of under- and malnourished children, women and men in the world. I have frequently pointed out that the future belongs to nations with grains and not guns. Our farmers are confronted with the challenge of producing food for 1.2 billion human beings and over 1 billion farm animals. The demoralising impact of the indifference shown to the safe storage of grains produced by hard labour in sun and rain by millions of farm women and men can only be imagined. The sense of national shame now prevailing because of the projection by media of the sad state of storage conditions should spur both Central and State governments into action. The storage can start in every village in the form of Grain Banks and rural godowns and extend to strategic locations (i.e., hunger hot spots) throughout the country. It is time we invested in a national grid of ultra-modern storage structures.

Rice, wheat and other grains can help to address protein-calorie undernutrition. But only attention to horticulture, milk and eggs can help to overcome hidden hunger caused by the deficiency of micro-nutrients like iodine, iron, zinc, vitamin A, vitamin B_{12}, etc. For example, drumstick or *Moringa* provides most of the micronutrients needed by the body. There are horticultural remedies for every nutritional malady and hence nutrition should be mainstreamed in the National Horticulture Mission. Home Science students well versed in nutrition should be inducted into the Mission. Like ASHAs (Accredited Social Health Activists) in the case of the National Rural Health Mission, they should become Nutrition Messengers in both rural and urban areas. A Shiksha Food Park (a Training Food Park) should be established by the Ministry of Food Processing in major Home Science Colleges to train women's self-help groups in the science and art of food processing and preservation. Attention to the preservation of perishable commodities is as important as attention to the safe storage of foodgrains.

Sow extensively during the *rabi* season

During the *rabi* season it will be prudent to review the arrangements for the supply of needed inputs like credit, insurance, seed, fertiliser and extension. Special efforts will have to be made to mount compensatory production programmes in areas affected by unfavourable weather during *kharif*. *Rabi* pulses and oilseeds need particular attention from the point of view of choice of variety, soil health enhancement and plant protection. The aim should be to achieve a higher per-day and per-crop productivity, so that even if there is a premature rise in night temperature, yields do not go down.

To sum up, we had paid considerable attention to grain storage during our "ship to mouth" existence days, as evident from the grain storage structures built in major ports. Home-grown grains have however failed to receive as much attention as the imported ones. Similarly, the Save Grain Campaign which was launched when we were food deficit was abandoned at a time when we needed it most. It is to be hoped that the prevailing widespread interest in saving and sharing grains will lead to an effective "distribute, procure, store and sow" movement. Without this pre-requisite, it will be difficult to implement a legal right to food for all.

❑❑❑

Chapter 12

Sustainable Food Security: Pre-requisites for Success

The President of India, in her address to both Houses of Parliament on 4 June 2009 announced:

> My Government proposes to enact a new law — the National Food Security Act — that will provide a statutory basis for a framework which assures food security for all. Every family below the poverty line in rural as well as urban areas will be entitled, by law, to 25 kilograms of rice or wheat per month at Rs. 3 per kilogram. This legislation will also be used to bring about broader systemic reform in the public distribution system.

Since then, various arms of government as well as civil society organisations have been working on methods of redeeming this pledge. The National Advisory Council (NAC) headed by Sonia Gandhi has recently provided a broad framework for achieving the goal of food for all and forever. The suggestions of NAC include the initiation as soon as possible of special programmes all over the country to insulate from hunger and malnutrition pregnant and nursing mothers, 0-3-years-old infants, street children, destitutes, old and infirm persons, patients suffering from HIV/AIDS, tuberculosis and leprosy, physically or mentally handicapped persons and other disadvantaged citizens. Such special nutrition support programmes may need about 10 million tonnes of foodgrains annually. NAC has stressed that in the design of the delivery system, there should be a proper match between challenge and response, as for example, the starting of community kitchens in urban areas for ensuring that destitutes, street children and old and infirm persons do not go to bed hungry. The highest priority in the special nutrition support programmes should go to pregnant women, in order to eliminate maternal and foetal undernutrition and thereby the primary cause of children being born with low birthweight.

For providing food security for all, NAC has proposed a phased programme of implementation of the goal of universal PDS, starting in 2011-12 with either one-fourth of the districts or blocks, and covering the whole country by 2015, on lines similar to that adopted in the implementation of the Mahatma Gandhi National Rural Employment Guarantee Programme. This will provide time to develop essential infrastructure like grain storage, Village Knowledge Centres and the issue of Household Entitlements Passbooks. NAC is also developing inputs for the proposed Food Security Act covering legal entitlements and enabling provisions based on the principle of common but differentiated entitlements, taking into account the urgent unmet needs of the economically and socially underprivileged sections of the Indian family.

Food is the first among the hierarchical needs of a human being. Therefore, the Food Security legislation will be the most significant among the laws enacted by the Parliament of Independent India. When enacted, the Food Security Act will mark the fulfillment of Mahatma Gandhi's call for a hunger-free India. Therefore, the proposed Food Security Act should lend itself for effective implementation, both in letter and spirit. This will call for serious attention to the following four pre-requisites. These are food production, procurement, preservation and public distribution. Let me explain what needs to be done in these areas.

Production

We face a formidable task on the food production front. Production should be adequate to provide balanced diets for over 1.2 billion persons. In addition, over a billion cattle, buffalo, goat, sheep, poultry and other farm animals need feed and fodder. The most urgent task is the implementation of the recommendations of the National Commission on Farmers (NCF) made in the five reports submitted to the Union Minister of Agriculture between 2004-06, and the National Policy for Farmers placed in Parliament in November 2007. The Policy for Farmers and the NCF reports provide a road map for strengthening the ecological and economic foundations for sustainable advances in productivity and production in the different agro-ecological zones of the country. They indicate methods of imparting an income orientation to farming, thereby helping to bridge the prevailing gap between potential and actual yields and income in most farming systems. Since land and water are shrinking resources for agriculture, and since climate change is a real threat, NCF has urged the spread of conservation and climate-resilient farming. Also, a conservation-cultivation-consumption-commerce chain should be promoted in every block. This will call for considerable technological and skill upgradation of farming practices. Methods of achieving a small farm management revolution resulting in higher productivity, profitability and stability under both irrigated and rain-fed conditions are described in the NCF reports.

The widening of the food basket through inclusion of nutritious millets, the mainstreaming of nutritional considerations in the National Horticulture Mission, and the consumption of salt fortified with iron and iodine will help to reduce both chronic protein-energy undernutrition and hidden hunger caused by the deficiency of micronutrients like iron, iodine, zinc, Vitamin A and Vitamin B_{12} in the diet. A sustainable food security system can be developed only with home-grown food, and not imports. Hence, we should mobilise all sources of nutrition available in the country.

Procurement

Procurement should cover not only wheat and rice, but also *jowar, bajra, ragi,* minor millets as well as pulses. When we started the High Yielding Varieties Programme in 1966, we included *jowar, bajra* and maize along with rice and wheat in order to keep the food basket wide. Community Grain Banks, operated under the social oversight of Gram Sabhas, will facilitate the purchase and storage of local grains. Farmers are now worried that government may be tempted to lower the minimum support price (MSP) to reduce the subsidy burden. This will be disastrous and will kill the food security system. MSP should be according to the NCF formula of C2 (i.e., total cost of production) plus 50 per cent. The actual procurement price should be fixed at the time of harvest, taking into account the escalation in the cost of inputs like diesel since the time the MSP was announced. It should be remembered that, unlike in developed countries where hardly 2 to 3 per cent are farmers, the majority of consumers (over 60 per cent) in our country are farmers. Their income security is hence vital for food security.

Preservation

Safe storage of the grains procured is the weakest link in the food security chain. While in Europe and North America as well as many developing countries, modern grain silos can be found in every farm, we are yet to develop a national grid of modern grain silos. Post-harvest losses are high both in foodgrains and in perishable commodities like vegetables and fruits. Although a Rural Godown Scheme was initiated in 1979, it is yet to take off in any significant manner. Meanwhile, government has closed down the "Save Grain" campaign a few years ago, thus ending one of the most relevant programmes in the context of food security.

Public Distribution

The strengths and weaknesses of the world's largest public distribution system operating in India are now being discussed widely. Corruption and leakages are widespread and should be ruthlessly put down. There are however States like Tamil Nadu, Kerala and Chattisgarh where PDS is being carried out in an efficient and corruption-free manner. The challenge is to learn from the successful models and convert the unique into universal.

In the ultimate analysis, what is relevant for human health and productivity is nutrition security at the level of each child, woman and man. We have to shift viewing food security at the aggregate level to ensuring nutrition security at the level of each individual. This will call for concurrent attention to food availability, access and absorption. Our agriculture is in a state of crisis, both from the economic and ecological points of view. Unless immediate attention is paid to soil health care and enhancement, water conservation and efficient use, adoption of climate-resilient technologies, timely supply of the needed inputs at affordable prices, credit and insurance and above all, producer-oriented marketing, our goal of achieving a higher growth rate in agriculture will not be realised. In the area of access, the Mahatma Gandhi National Rural Employment Guarantee Programme and the proposed Food Security Act assuring 35 kg of staple grains at Rs. 3 per kg will help. We have to combine this with additional efforts to create avenues for market-driven non-farm enterprises. When China started its agricultural reform, a two pronged strategy was adopted — higher productivity and profitability of small farms and greater opportunities for non-farm employment and income through Township Village Enterprises. Unfortunately we still see a gross mismatch between production and post-harvest technologies, resulting in the spoilage of foodgrains and missed opportunities for value addition and agro-processing. In particular, our use of agricultural biomass is generally wasteful and does not lead to the creation of new jobs and income.

In the field of absorption of food in the body, it is important to pay attention to clean drinking water, sanitation and primary health care. Even in a State like Tamil Nadu where steps have been taken for ensuring food at affordable cost, the food insecurity analysis of MSSRF (carried out along with the World Food Programme) shows that food security is far better in households with toilets. The Rajiv Gandhi Drinking Water Mission, the Total Sanitation Programme and the National Rural Health Mission are all important for food security. Our global rank in the areas of poverty and malnutrition will continue to remain unenviable, so long as we do not enable all our citizens to have a productive and healthy life. The proposed Food Security Act provides the last chance for saving nearly 40 per cent of our population from the hunger trap.

❑❑❑

Chapter 13

Pathway for Food Security for All

Pranab Mukherjee in his budget speech delivered on 26 February 2010 announced: 'We are now ready with the draft Food Security Bill which will be placed in the public domain very soon.' Although no official draft has been placed on the website of the concerned Ministry, several leading organisations and individuals have questioned the adequacy of the steps proposed to be taken under the Bill for achieving the goal of a hunger-free India. Based on Article 21 of the Constitution, the Supreme Court of India has rightly regarded the right to food as a fundamental requirement for the right to life. Many steps have been taken since Independence to adopt Mahatma Gandhi's advice for an *antyodaya* approach to hunger elimination. In spite of numerous measures and social safety net programmes, the number of undernourished persons has increased from about 210 million in 1990-92 to 252 million in 2004-06. About half of the world's under-nourished children are in India. Also, there has been a general decline in per capita calorie consumption in recent decades. Grain mountains and hungry millions continue to co-exist.

We are fortunately moving away from a patronage to a rights approach in areas relating to human development and well-being. The Acts relating to the Right to Information, Education, Land for Scheduled Tribes and Forest

Dwellers, and Rural Employment are examples of this trend. The proposed Food Security Bill, when enacted, will become the most important step taken since 1947 in launching a frontal attack on poverty-induced endemic hunger in the country. The adverse impact of undernutrition on human health and productivity is well known. In addition to denying every citizen the right to a productive and healthy life, chronic undernutrition also makes it difficult to overcome diseases like tuberculosis, leprosy and HIV/AIDS. In the case of such diseases, a food–cum-drug approach is needed to ensure success of the treatment.

The numerous programmes introduced by the Government of India from time to time for improving the nutritional status of children, women and men include the following:

- *Ministry of Women and Child Development*: Integrated Child Development Services (ICDS), Kishori Shakti Yojana, Nutrition Programme for Adolescent Girls, Rajiv Gandhi Scheme for Empowerment of Adolescent Girls
- *Ministry of Human Resource Development*: Mid-day Meals Scheme, Sarva Shiksha Abhiyan
- *Ministry of Health and Family Welfare*: National Rural Health Mission, National Urban Health Mission
- *Ministry of Agriculture*: Rashtriya Krishi Vikas Yojana, National Food Security Mission, National Horticulture Mission
- *Ministry of Rural Development*: Rajiv Gandhi Drinking Water Mission, Total Sanitation Campaign, Swarna Jayanti Gram Swarojgar Yojana, Mahatma Gandhi National Rural Employment Guarantee Programme
- *Ministry of Food*: Targeted Public Distribution System, Antyodaya Anna Yojana, Annapoorna

Notwithstanding this impressive list of interventions, the present situation in the field of child nutrition is that 42.5 per cent of children below 5 years of age are underweight and 40 per cent of children below 3 years are undernourished.

In order to succeed in ensuring food security for all, we should be clear about the definition of the problem, the precise index of measuring impact and a road map for achieving the goal. Presently, the discussion is mainly around the definition of poverty and on methods of identifying the poor. We have the most austerely defined poverty line in the world and the official approach appears to be to restrict support only to BPL (below poverty line) families. Calculations of the numbers of BPL families (taking 4 persons as the average size of a family), the number varies from 9.25 crore (Suresh Tendulkar Committee) to 20 crore (Justice D. P. Wadhwa).

Food security, as internationally understood, involves physical, economic and social access to a balanced diet, safe drinking water, environmental hygiene and primary health care. Such a definition will involve concurrent attention to the availability of food in the market, the ability to buy the needed food and the capability to absorb and utilise the food in the body. Thus, food and non-food factors (i.e., drinking water, environmental hygiene and primary health care) are involved in food security.

As earlier mentioned, we have numerous Central Government schemes dealing with nutrition support, drinking water, sanitation and health care. In addition, most State Governments have additional schemes such as extending support to mothers to feed newborn babies with mothers' milk for at least the first six months. States like Tamil Nadu and Kerala have universal public distribution system. Unfortunately, the governance of the delivery of such programmes is fragmented and a "deliver as one" approach is missing. Also, a lifecycle approach in the development and delivery of nutrition support programmes, starting with pregnant women and ending with old and infirm persons, is lacking. Our current unenviable status in the world in the field of nutrition is largely because of the absence of a good governance system which can measure outlay and output in an unbiased manner. Therefore, more than new schemes, the governance of existing schemes needs attention.

Common and Differentiated Entitlements

In my view, the National Food Security Bill should be so structured that it provides common and differentiated entitlements. The common entitlements should be available to every citizen of the country. This should include a universal PDS, clean drinking water, sanitation, hygienic toilets, and primary health care. The differentiated entitlements could be restricted to those who are economically or physically handicapped. Such families can be provided with wheat or rice in the quantity decided at Rs.3 per kg, as is now being proposed. The availability of cheap staple grains will only help BPL families address the problem of access to food at affordable prices, but not give them economic access to balanced diets. At the prevailing price of pulses, such families will find protein-rich foods out of reach. Similarly, hidden hunger caused by micro-nutrient deficiencies like iron, iodine, zinc, vitamin A and Vitamin B_{12} will continue to persist. The question thus arises as to what we want to achieve from the proposed Food Security Bill. Should the Bill enable every child, woman and man to have an opportunity for a healthy and productive life, or just have access to the calories required for existence ? If the aim is the latter, the title "Food Security Bill" will be inappropriate.

We can learn much from the "Zero Hunger" programme of Brazil, where a holistic view of food security has been adopted. The measures include steps to enhance the productivity of smallholdings as well as the consumption capacity of the poor. Our farmers will produce more if we are able to purchase more. Emphasis on agricultural production, particularly small farm productivity, will as a single step make the largest contribution to poverty eradication and hunger elimination. While universal PDS should be a legal entitlement, the other common entitlements could be indicated in the Bill for the purpose of monitoring and integrated delivery. The involvement of Gram Sabhas and Nagarpalikas in monitoring the delivery systems will help to improve efficiency and curb corruption.

What is desirable should also be implementable. The greatest challenge in implementing the common and differentiated food entitlements under the Bill will be the production of adequate quantities of staple grains. Fortunately, the untapped production reservoir, even with the technologies now on the shelf, is high in both irrigated and rain-fed farming systems. Doubling the production of rice and wheat in eastern India and pulses and oilseeds in rain-fed areas during this decade is feasible. The 2010-11 budget indicates measures for initiating a "bridge the yield gap movement" in eastern India, and for stimulating a pulses and oilseeds revolution through the organisation of 60,000 Pulses and Oilseed Villages where concurrent attention will be given to the conservation of soil and water, cultivation of the best available strains, consumption of local grains (like *jowar, bajra, ragi,* etc.) and commerce at prices fair to farmers. National and State efforts should be supported by efforts at the local-body level for building a community food security system involving seed, grain and water banks.

The National Commission on Farmers (2006) in its comprehensive recommendations on building a sustainable national nutrition security system had calculated that about 60 million tonnes of foodgrains will be needed for universal PDS. The differentiated entitlements for BPL families for foodgrains at low cost will involve only additional cash expenditure. In fact, food stocks with Government may touch 60 million tonnes by June 2010.

For the Government to remain at the commanding height of such a food security system combining universal and unique entitlements, the four-pronged strategy indicated in Pranab Mukherjee's budget speech should be implemented jointly by panchayats, State governments and Union Ministries with speed and earnestness. Just as the Golden Quadrilateral initiative of Atal Behari Vajpaye electrified the national road communication infrastructure, we need a Golden Quadrilateral in the development of a national grid of modern grain storages. Will Manmohan Singh leave his footprints on the sands of time in the case of safe storage of foodgrains and perishable commodities all over the country as an essential requirement for food security, as Vajpaye has done in the case of roads ?

I hope we will not lose this historic opportunity for ensuring that our nation takes to a development pathway which regards the nutrition, health and well-being of every citizen as the primary purpose of a democratic system of governance.

❑❑❑

Chapter 14

Biodiversity and Poverty Alleviation

Biodiversity provides the building blocks for sustainable food, health and livelihood security systems. It is the feedstock for both the biotechnology industry and a climate-resilient farming system. Because of its importance for human well-being and survival, a Convention on Biological Diversity (CBD) was adopted at the UN Conference on Environment and Development held at Rio de Janeiro in 1992. The Convention's triple goals are conservation, sustainable use, and equitable sharing of benefits. CBD defines biological diversity as follows:

> Variability among living organisms from all sources including, inter alia, terrestrial, marine and other aquatic ecosystems and the ecological complexes of which they are part; this includes diversity within species, between species and of ecosystems.

The Convention stresses the need for respecting, preserving and maintaining the knowledge, innovations, and practices of indigenous and local communities embodying traditional lifestyles relevant for the conservation and sustainable use of biological diversity. It also calls for the promotion of their wider application with the approval and involvement of the holders of such knowledge, innovations, and practices and for the equitable sharing of the benefits arising from the utilisation of such knowledge, innovations, and practices (Article 8J).

The Convention also recognises that the biodiversity existing within a country is the sovereign property of its people. India is a signatory to CBD and has enacted a National Biodiversity Act which has been in force since 2002. India is classified as a mega biodiversity area from the point of view of species richness and agro-biodiversity. However, two of the major biodiversity-rich areas, north-east India and the Western Ghats region, are also classified as "hot spot" areas from the point of view of threats to biodiversity.

In spite of the importance given to the conservation of biodiversity, genetic erosion is progressing in an unabated manner, both globally and nationally. For example, 12 per cent of birds, 21 per cent of mammals, 30 per cent of amphibians, 27 per cent of coral reefs and 35 per cent of conifers and cycads are currently facing extinction. According to the World Conservation Union (IUCN), over 47,677 species may soon disappear. A comprehensive study published in *Science* (29 April 2010) has revealed that there has been no notable decrease in the rate of biodiversity loss between 1970 and 2010. Even a very unique species like the orang-utan, the closest relative of man, is threatened with extinction in the island of Borneo.

To generate awareness of the urgency of genetic resources conservation, 22 May of every year is being commemorated as the International Day for Biological Diversity, and 2010 as the International Year of Biodiversity. The challenge now is for every country to develop an implementable strategy for saving rare, endangered and threatened species through education, social mobilisation and regulation. Meaningful results can be obtained only if biodiversity conservation is considered in the context of sustainable development and poverty alleviation. Indira Gandhi pointed out at the UN Conference on the Human Environment held at Stockholm in 1972 that unless we attend concurrently to the needs of the poor and of the environment, the task of saving our environmental assets will not be easy. Biodiversity loss is predominantly related to habitat destruction largely for commercial exploitation as well as for alternative uses like roads, buildings, etc. Invasive alien species and unsustainable development are other important causes of genetic erosion. How can we reverse the paradigm and enlist development as an effective instrument for conserving biodiversity? Let me cite a few examples to illustrate how biodiversity conservation and development can become mutually reinforcing.

In 1990, I visited MGR Nagar, a village near Pichavaram in Tamil Nadu, for studying the mangrove forests of that area. The families living in MGR Nagar were extremely poor and were still waiting for the benefits of government schemes to be allocated to them. The children had no opportunities for education, the schools were far away and getting admission was difficult. I then told my colleagues that saving mangrove forests without saving the children for whose well-being these forests were being saved made no sense. With the help of a few donors, we started a primary school in the village for all the children, irrespective of their age. A few years later, the State government took over the school and expanded its facilities. It now needs to be upgraded into a higher secondary school. After the 2004 tsunami, the hutment dwellings have been replaced by brick houses and the whole scenario of MGR Nagar has totally changed. During the tsunami, the mangroves served as speed breakers and saved the people of the village from the fury of the tidal waves. Everyone in the village now understands the symbiotic relationship between mangroves and coastal communities, that the root exudate from the mangrove trees enriches the water with nutrients and promotes sustainable fisheries. It is clear that hereafter mangroves in this region will be in safe hands.

Another example relates to the tribal families of Kolli Hills in Tamil Nadu. The local tribal population had been cultivating and conserving a wide range of millets and medicinal plants. However, due to lack of market demand for traditional foods, they had to shift to more remunerative crops like tapioca and pineapple. The millet crops cultivated and consumed by them for centuries were rich in protein and micronutrients. They were also much more climate resilient, since mixed cropping of millets and legumes minimises risks arising from unfavourable rainfall. Such risk distribution agronomy is the saviour of food security in an era of climate change. How then can we revitalise the conservation traditions of tribal families, without compromising on their economic well-being ? MSSRF scientists started a programme designed to create an economic stake in conservation, by both value addition to primary products and by finding niche markets for their traditional foodgrains. Commercialisation thus became the trigger for conservation. Today many of the traditional millets are once again being grown and consumed. They now proudly sing, "Biodiversity is our life", which is also the key message of the International Year for Biodiversity.

A third example relates to the tribal areas of the Koraput region of Odisha (formerly Orissa), which is an important centre of diversity of rice. Fifty years ago, there were over 3500 varieties of rice in this area. Today this number has come down to about 300. Even with these three hundred varieties, it is essential that the tribal families derive some economic benefit from the preservation of such rich genetic variability in rice. Now, they, in partnership with scientists, have developed improved varieties like *Kalinga Kalajeera*, which fetches a premium price in the market. For too long, tribal and rural families have been conserving genetic resources for public good at personal cost. It is time that we recognise the importance of promoting a genetic conservation continuum, starting with the simple situation of *in situ* on-farm conservation of landraces by local communities, and extending to technological breakthroughs like the preservation of a sample of genetic variability under permafrost conditions at locations like Svalbard near the North Pole maintained by the Government of Norway or Chang La in Ladakh where our Defence Research and Development Organisation has established a conservation facility.

While giving operational content to the concept of sustainable development, we should ask some of the questions Dudley Sears asked decades ago:[1]

> The questions to ask about a country's development are: What has been happening to poverty ? What has been happening to unemployment ? What has been happening to inequality ? If all three of these have become less severe, then beyond doubt this has been a period of development for the country concerned. If one or two of these central problems have been growing worse, especially if all three have, it would be strange to call the result "development", even if per capita income doubled.

In dealing with issues relating to biodiversity, we should also ask similar questions. Is the biodiversity management system conducive to the reduction of poverty, promotion of gender equity, and the generation of livelihood opportunities ? The need for gender and social equity in sharing benefits from the commercial use of biodiversity cannot be overemphasised, if we

---

[1] Sears, D, 1969. "The Meaning of Development". *International Development Review* 11(4)

are to succeed in preventing further genetic erosion. Biodiversity rich countries are characterised by agroecological variability. In addition, there is a strong positive correlation between cultural diversity and agro-biodiversity. Women, in particular, tend to conserve and improve plants of value in strengthening household nutrition and health security. It is, therefore, imperative to give explicit recognition to the role of women in genetic resource conservation and enhancement.

How can we harness biodiversity for poverty alleviation ? Obviously, this can be done only if we can convert biodiversity into jobs and income on a sustainable basis. Several institutional mechanisms have been developed at MSSRF for this purpose, such as biovillages and biovalleys. In biovillages, the conservation and enhancement of natural resources like land, water and biodiversity become priority tasks. At the same time, the biovillage community aims to increase the productivity and profitability of small farms and create new livelihood opportunities in the non-farm sector. Habitat conservation is vital for preventing genetic erosion. In a biovalley, the local communities try to link biodiversity, biotechnology and business in a mutually reinforcing manner. For example, the herbal biovalley under development in Koraput aims to conserve medicinal plants and local foods and convert them into value-added products based on assured and remunerative market linkages. Tribal families in Koraput have formed a "Biohappiness Society".

There is need for launching a Biodiversity Literacy Movement, so that right from childhood everyone is aware of the importance of diversity for the maintenance of food, water, health and livelihood security as well as a climate-resilient food production system. The Government of India has started programmes like DNA and Genome Clubs to sensitise schoolchildren about the importance of conserving biodiversity. The Government has also started recognising and rewarding the contributions of rural and tribal families in the field of genetic resources conservation through Genome Saviour Awards. We need similar awards for those who are conserving breeds of animals, forests and fishes.

The Biodiversity Act envisages action at three levels: the Panchayat Biodiversity Committee (responsible for conservation as well as for operationalising the concept of prior informed consent and benefit sharing), the State Biodiversity Board and the National Biodiversity Authority. These three units of the bioresources conservation movement should ensure that all development programmes are subjected to a biodiversity impact analysis, so that economic advance is not linked to biodiversity loss. The Biodiversity Day and the Biodiversity Year remind us that we should do everything possible to spread bioliteracy among the public and usher in an era of biohappiness in biodiversity rich areas. Then, "biodiversity hot spots" will become "biodiversity happy spots".

❑❑❑

Chapter 15

Biodiversity and Sustainable Food Security

Demographic explosion, environment pollution, habitat destruction, enlarging ecological footprint, widespread hunger and unsustainable lifestyles, and potential adverse changes in climate all threaten the future of human food, water, health and livelihood security systems. 2010 appears to mark the beginning of uncertain weather patterns and extreme climate behaviour. Events like temperature rise, drought, flood, coastal storms and rise in sea level are likely to present new challenges to the public, professionals and policy makers. Biodiversity has so far served as the feedstock for sustainable food and health security and can play a similar role in the development of climate-resilient farming and livelihood systems. Biodiversity is also the feedstock for the biotechnology industry. Unfortunately, genetic erosion and species extinction are now occurring at an accelerated pace due to habitat destruction, alien species invasion and spread of agricultural systems characterised by genetic homogeneity. Genetic homogeneity increases genetic vulnerability to biotic and abiotic stresses. To generate widespread interest in biodiversity conservation, the UN General Assembly has declared 2010 as the International Year of Biodiversity.

The Global Convention on Biodiversity (CBD) adopted at the UN Conference on Environment and Development held at Rio de Janeiro in

1992, and the International Treaty on Plant Genetic Resources for Food and Agriculture adopted by Member Nations of FAO in 2001 provide a road map for the conservation and sustainable and equitable use of biodiversity. CBD emphasises that biodiversity occurring within a nation is the sovereign property of its people. Hence, the primary responsibility for conserving biodiversity, using it sustainably and equitably, and preserving it for posterity rests with each individual nation. This implies that all nations should subject development programmes to a Biodiversity Impact Analysis in order to ensure that economic advance is not linked to biodiversity loss. Inter-generational equity demands that we must preserve for posterity at least a representative sample of the biodiversity existing in our planet today.

Initiatives like the recognition of Globally Important Agricultural Heritage Sites of FAO and the World Heritage Sites of UNESCO are important to generate interest in the conservation and enrichment of unique biodiversity sites. Particular attention will have to be given to sustaining the protected areas through public education and social mobilisation, in addition to appropriate regulation. Unfortunately, many of the protected areas, national parks and biosphere reserves are facing serious anthropogenic pressures. Based on the model of the Biosphere Trust for the conservation of the Gulf of Mannar Biosphere Reserve in India developed by MSSRF, biosphere reserves could be jointly managed by local communities and government departments. The concept of participatory forest management should be extended to national parks and biosphere reserves. This will help to foster among the local population the feeling that they are trustees of these unique gifts of nature.

Special attention should be paid to biodiversity hot spots. Through public cooperation, they should be converted into biodiversity "happy spots", where the sustainable use of biodiversity helps to generate new jobs and income. Coastal biodiversity has not received adequate attention. Mangrove wetlands are under various degrees of degradation. The Joint Mangrove Forest Management procedure developed by MSSRF should be implemented wherever mangrove genetic resources still occur. Infrastructure for strengthening community conservation like drying yards, seed storage and seed testing facilities needs to be supported in all agro-biodiversity hot spots.

Biodiversity conservation and sustainable management should become a national ethic. Government agencies including local self-government authorities like panchayats could play an important role in both spreading biodiversity literacy through Community Biodiversity Registers and by creating the necessary infrastructure like Gene and Seed Banks. Awareness of the relationship between biodiversity and human health and farm animal survival should become widespread. Special training programmes should be organised to enable panchayat committees to become well versed with the provisions of the Biodiversity and Protection of Plant Varieties and Farmers' Rights Acts, particularly with those relating to prior informed consent, access and benefit sharing as well as the gene and biodiversity funds.

Women play a lead role in biodiversity conservation and sustainable use. Mainstreaming the gender dimension in all conservation and food security programmes is a must. Women conservers should be enabled to continue their conservation ethos, by providing support for essential infrastructure like seed storages. Agro-biodiversity is the result of interaction between cultural diversity and biodiversity. An important aspect of cultural diversity is culinary diversity. Every step should be taken to recognise and preserve cultural diversity and to blend traditional wisdom with modern science.

The role of farmers and farming in the mitigation of climate change has not so far been adequately recognised and appreciated. Farmers can help build soil carbon banks and at the same time improve soil fertility through fertiliser trees. Mangrove forests are very efficient in carbon sequestration. Biogas plants can help to convert methane emissions into energy for the household. Hence, a movement should be started at global, national and local levels for enabling all farmers with small holdings and a few farm animals to develop a water-harvesting pond, plant a few fertiliser trees and establish a biogas plant in their farms. I reiterate: just these — a farm pond, some fertiliser trees and a biogas plant — will make every small farm contribute to climate change mitigation, soil health enhancement and water for a crop life-saving irrigation.

To strengthen the linkages between biodiversity and food security, there is need to enlarge the food basket by including a wide range of millets, tubers and legumes in the diet. Nutrition security can be strengthened by introducing horticultural remedies for nutritional maladies, like the deficiency of iron, iodine, zinc, vitamin A, vitamin B_{12} and other micronutrients in the diet. Biodiversity in medicinal plants helps to strengthen health security. The role of biodiversity in sustaining livelihoods can be enhanced through crop-livestock-fish integrated farming systems. Bio-resources should be converted into jobs and income meaningful to the poor in an environmentally sustainable manner.

Kerala, a case study

I would like to describe here the State of Kerala in south-west India which is an agro-biodiversity paradise. Kerala is rich in crops like rice, banana, jackfruit, tubers, spices, medicinal plants, coconut, plantation crops, coastal halophytes, inland and marine fishes, large and small ruminants including the Vechur cow and Malabari and Attapadi goats. The medicinal plant wealth has helped Kerala to perfect the science of ayurveda and thereby become a preferred state for health tourism. The challenge now lies in both preserving and enriching this biological wealth and in converting bio-resources into jobs and income on a sustainable basis.

Kerala has a long tradition of *in situ* on-farm conservation in crops like rice, spices and tubers as well as *ex situ* preservation through sacred groves, botanical gardens, biosphere reserves and aquaria. Tribal communities have conserved life-saving crops particularly tubers and medicinal plants, and traditional healers have deep knowledge of the therapeutic value of local flora. Speciality rices like *Njavara* have been identified and conserved. Farmers have been serving as conservers, breeders and cultivators. Several important varieties like *Njallani* in cardamom have been developed by farmers in Idukki district. Kuttanad farmers have perfected the art and science of growing paddy below sea level. This knowledge will be of immense value in protecting coastal agriculture in the event of a rise in sea level.

Climate change presents mega-threats to Kerala's food and water security systems as well as to the lives and livelihoods of coastal communities. Sea level rise will cause serious threats to coastal ecosystems as well as to coastal mineral wealth, as, for example, the monozite and thorium deposits. A temperature rise of 2 degree Celsius will affect the production and productivity of plantation crops like coffee, tea, spices and rubber, in addition to annual crops like rice. Change in precipitation may cause both drought and floods as well as soil erosion and decrease in soil fertility. The forest biodiversity and medicinal plant wealth of Kerala will also be adversely affected. Ecosystem services will be disrupted. Vector-borne diseases will affect plant, animal and human health. Kerala may experience a large influx of "climate refugees" from coastal to inland areas. To prepare for such threats, both anticipatory research using advanced technologies as well as participatory research with local communities including tribal families will be needed so that coping mechanisms combining frontier science and traditional wisdom can be developed and put in place soon.

Kerala has the unique advantage of becoming a world leader in managing the consequences of sea level rise. The Kuttanad area may be declared as a Special Agriculture Zone as it is the only region in India with experience of cultivating rice under below sea level conditions; it is a Ramsar site; it is a unique wetland promoting rice-fish rotation; it is an area of thriving water tourism; it is a biodiversity paradise in flora and fauna, and it provides uncommon opportunities for learning how to manage the impact of sea level rise. An International Research and Training Centre for Below Sea Level Farming should be established in Kuttanad.

The area extending from the Silent Valley Biosphere Reserve to Wayanad may be developed as a herbal biovalley, to promote the conservation and sustainable and equitable use of the genetic diversity occurring in medicinal plants. Micro-enterprises supported by micro-credit may be organised by women's SHGs along the biovalley. All rare, endangered and threatened (RET) species in the biovalley should be protected and multiplied. The products of the biovalley may be given a brand name. Conservation and commercialisation will then become mutually reinforcing, and there will be an economic stake in conservation. Today, there is an economic interest in the unsustainable exploitation of medicinal plant resources, and this needs to be halted and reversed through the medicinal plants biovalley.

Suggestions for the National Action Plan of the Biodiversity Authority of India

I would like to set out some of the steps we should take for making the conservation of biodiversity everybody's business.

Deliver as one

The importance of the use and conservation of biodiversity in agro-ecosystems should be recognised in national development plans (including poverty reduction programmes). This necessitates integration of approaches across government departments confronting rural development, food security, poverty reduction, environment and climate change. To the extent feasible the "deliver as one" approach should be adopted, in order to achieve convergence and synergy among different ongoing programmes.

Build partnerships

Effective use of agro-biodiversity is the key to realising its development impact and its conservation. This requires development of markets for products of diverse agriculture, especially underutilised crops, different animal genetic breeds, etc. This can be built on public-private partnerships and development of agribusinesses benefiting rural communities.

Strengthen the role of farming and tribal communities

Farming and tribal communities have a major role in delivering the benefits of agrobiodiversity. The following measures should be incorporated:

- Integrate community *in situ* and *ex situ* conservation in the national biodiversity conservation strategy. *In situ* conservation will start from the field. *Ex situ* conservation can take the form of sacred groves and heritage trees, as well as botanical and zoological gardens.
- Organise field Gene Banks, Seed anks and Grain Banks at the local level. This will help to promote *in situ* on-farm conservation of landraces and enlarge the food basket and thereby strengthen local level crop and food security.
- Establish special Gene Banks for climate-resilient crops.
- Recognise and reward primary conservers of biodiversity through initiatives like the Genome Saviour Award instituted by the Plant Variety Protection and Farmers' Rights Authority of India.

Conservation science

R&D priorities should be re-focused to enhance the productivity of bio-diverse agriculture, including the need to optimise genetic diversity (plants and animals). For example, landraces and wild crop relatives should be characterised, evaluated and utilised in crop improvement programmes to transfer traits relevant to climate change, such as drought and heat resistance and flood and salinity tolerance. Breeding for per-day yield rather than per-crop productivity should receive priority. These concerns should be reflected in higher education curricula and research agenda.

Climate-resilient farming systems

Climate change will demand modifications to farming systems (e.g., cultivars, land use, water use management, animal selection) and increased environmental risk management. This will require prioritisation of the social and agro-ecological zones most at risk. Biodiversity is the feedstock not only for food and health security, but also for the management of climate change. Gene Banks for a warming planet have become urgent as an essential element of climate-resilient farming systems. The prospects for climate change have added urgency to efforts designed to save every gene and species now existing on our planet. The initiative of the Government of Norway in establishing a Global Seed Vault at Svalbard, and of the Defence Research and Development Organisation of the Government of India in establishing a similar facility under perma-frost conditions at Chang La in the Himalayas are welcome steps.

Land-use patterns

Since land use represents a third of global greenhouse gas emissions, this must be reversed. Farmers can help build soil carbon reserves and at the same time improve soil fertility through fertiliser trees and conservation agriculture. Mangrove forests are very efficient in carbon sequestration and should be protected.

Economic value of ecosystem services

According economic value to ecosystem services like land, water, biodiversity and climate and putting in place mechanisms for payment for such services will help to reduce the ecological footprint and thereby help to achieve a balance between biocapacity and natural resources exploitation.

Biodiversity literacy

An extensive and well-designed biodiversity awareness and literacy campaign should be launched, starting with schoolchildren and extending up to the adult population. Such a biodiversity literacy programme must involve the integrated use of traditional and new media. Village Knowledge Centres could be utilised for sensitising the local population on the threats to biodiversity and the names and locations of the rare, endangered and threatened species occurring in that area. University students and civil society organisations can be assisted in saving RET species. The preparation of local level Biodiversity Registers can be promoted. Genetic gardens can be promoted in schools and colleges. There is need for promoting among the younger generation an awareness of the vital significance of agrobiodiversity for the well-being of the future generations. Genome Clubs designed to promote genetic and biodiversity literacy may be organised in all schools and colleges.

Climate care movement

The greatest casualty of climate change will be food and water security. Biodiversity helps to mitigate the adverse impact of climate change.

A climate care movement at the local, national and global levels must pay specific attention to the following:

- Gene care and conservation
- Climate literacy
- Appointment of local-level Community Climate Risk Managers
- Promotion of appropriate mitigation and adaptation measures

In order to promote coordinated and concerted efforts in agrobiodiversity conservation and enhancement, it will be useful to constitute a Consortium for Agro-biodiversity Conservation and Enhancement in every State with members drawn from the government, academic, civil society, media and private sectors. Such a consortium should promote the conservation of germplasm of crops, farm animals, fisheries and forest trees. The consortium should help in ushering in an era of biohappiness arising from the sustainable use of bio-resources from creating more jobs and income.

To sum up, biodiversity is the prime mover of an ever-green revolution movement in agriculture, and the goal of achieving "food for all and for ever" is wholly dependent on its conservation and sustainable and equitable use. Mahatma Gandhi's plea that 'we should live in harmony with Nature and with each other' should guide community efforts in the management of biodiversity. The declaration by the United Nations that 2010 be observed as the International Year of Biodiversity is to remind humankind that biodiversity is the foundation for global food security and that conservation of biodiversity and natural resources should become a non-negotiable human ethic.

❑❑❑

Chapter 16

Priorities in Agricultural Research and Education

It was in 1968 that the term Green Revolution was coined by Dr. William Gaud of USA to describe advances in agriculture arising from productivity improvement. Even in 1968, I concluded that if farm ecology and economics go wrong, nothing else will go right in agriculture. I expressed my views in the following words in my lecture at the Indian Science Congress Session held at Varanasi in January 1968:

> Exploitive agriculture offers great dangers if carried out with only an immediate profit or production motive. The emerging exploitive farming community in India should become aware of this. Intensive cultivation of land without conservation of soil fertility and soil structure would lead, ultimately, to the springing up of deserts. Irrigation without arrangements for drainage would result in soils getting alkaline or saline. Indiscriminate use of pesticides, fungicides and herbicides could cause adverse changes in biological balance as well as lead to an increase in the incidence of cancer and other diseases, through the toxic residues present in the grains or other edible parts. Unscientific tapping of underground water will lead to the rapid exhaustion of this wonderful capital resource left to us through ages of natural farming. The rapid replacement of numerous locally-adapted varieties with one

> or two high-yielding strains in large contiguous areas would result in the spread of serious diseases capable of wiping out entire crops, as happened prior to the Irish potato famine of 1854 and the Bengal rice famine in 1942. Therefore, the initiation of exploitive agriculture without a proper understanding of the various consequences of every one of the changes introduced into traditional agriculture, and without first building up a proper scientific and training base to sustain it, may only lead us, in the long run, into an era of agricultural disaster rather than one of agricultural prosperity.

The above analysis led me to coin the term "ever-green revolution" to describe the enhancement of productivity in perpetuity without associated ecological harm. The pathways to an ever-green revolution are either organic farming or green agriculture. Green agriculture involves the adoption of environment friendly practices like integrated natural resources management and integrated pest management. It is our sacred duty to conserve and enhance the ecological foundations such as soil, water and biodiversity essential for sustained advances in agricultural productivity and profitability.

The present decade may mark the beginning of a new climate era, characterised by extreme and often unpredictable weather conditions and rise in sea levels. The recent Climate Conference in Copenhagen unfortunately failed to get a global commitment to halt economic growth based on high carbon intensity. The Climate Conference due to be held in Mexico in December 2010 will probably generate the political commitment essential to restrict the rise in global mean temperature to not more than 2°C, as compared to the mean temperature of today. Even a 2°C rise will adversely affect crop yields in South Asia and sub-Saharan Africa, which already have a high degree of prevalence of endemic hunger. It will also lead to the possibility of small islands getting submerged. The greatest casualty of climate change will be food, water and livelihood security. Farmers of the world can help to avoid serious famines by developing and adopting climate-resilient farming systems. 2010 has been declared by the United Nations as the International Year of Biodiversity. As I have said earlier, biodiversity is the feedstock for a climate-resilient agriculture. We should therefore redouble our efforts to prevent genetic erosion and to promote the conservation and sustainable and equitable use of biodiversity.

2010 will also witness a major conference at the United Nations Headquarters in New York to review the progress made since the year 2000 in achieving the UN Millennium Development Goals. The first among these goals is reducing hunger and poverty by half by 2015. Unfortunately the number of hungry children, women and men, which was 800 million in 2000, is now over a billion. This is partly due to a rise in food prices, thereby making it difficult for the poor to have access to balanced diets at affordable prices.

Adaptation to climate change

A group of scientists led by MSSRF have undertaken studies during the last five years in Rajasthan and Andhra Pradesh on climate change adaptation measures. The districts chosen were Udaipur in Rajasthan and Mehabubnagar in Andhra Pradesh. The approach adopted was to bring about a blend of traditional wisdom and modern science. The participatory research and knowledge management systems adopted under this programme during the past five years have provided many useful insights for developing a climate-resilient farming and livelihood security system. Five of the meaningful adaptation interventions have been the following:

- Water conservation and sustainable and equitable use: Families in the desert regions of Rajasthan have long experience in harvesting every drop of rainwater and using it economically and efficiently both for domestic and agricultural use. The traditional methods were reinforced with modern scientific knowledge, like the gravity flow method of water management.
- Promoting fodder security: Livestock and livelihoods are intimately related in arid and semi-arid areas. The ownership of livestock is also more egalitarian. The sustainable management of common property resources, particularly pasture land, is essential for ensuring fodder security. Therefore, high priority was given to the regeneration of pasture land and the equitable use of grazing land.
- More crop and income per drop of water: In areas where water for irrigation is the constraint, it is important that agronomic techniques which can help to increase yield and income per drop of water are standardised and popularised. One such method introduced under this project is the System of Rice Intensification (SRI). SRI was popularised in Andhra Pradesh, since this system of water and crop management helps to reduce irrigation water requirement by 30 to

40 per cent. This method thus helps to avoid the unsustainable exploitation of the aquifer.

- Weather information for all and climate literacy: What farmers need is location- specific meteorological information at the right time and place. Generic weather data will have to be converted into location-specific meteorological advice. For this purpose, mini-agro-meteorological stations managed by the local community were established. This has helped to impart climate literacy as related to food, water and livelihood security.
- Strengthening community institutions: Effective implementation of adaptation measures will need active group cooperation and community participation. Steps were taken to involve the grass-root democratic institutions like panchayats and Gram Sabhas. Also, Smart Farmers' Clubs were organised to give the power of scale in water harvesting, soil health management and other adaptation measures undertaken by farmers with smallholdings.

These interventions were supported by training, skill development, education and social mobilisation. A Training Manual was prepared by MSSRF for training one woman and one male member of every panchayat in the art and science of managing weather abnormalities, making them local-level Climate Rick Managers.

The work has highlighted the need for location-specific adaptation measures and for participatory research and knowledge management. The adaptation interventions have also highlighted the need for mainstreaming gender considerations in all interventions. Women will suffer more from climate change, since they have been traditionally in charge of collecting water, fodder and fuel wood, and have been shouldering the responsibility for the care of farm animals as well as for post-harvest technology. All interventions should therefore be pro-nature, pro-poor and pro-women.

The last five years have been an extremely rewarding learning period. The results and experience have shed light on the way forward. It is clear that to promote location-specific and farmer-centric adaptation measures, India will need a Climate Risk Management Research and Extension Centre at each of the 127 agro-ecological regions in the country. Such centres should prepare Drought, Flood and Good Weather Codes that can help to minimise the adverse impact of abnormal weather and to maximise the benefits of favourable monsoons and temperature. Risk surveillance and early warning should be the other responsibilities of such centres.

The work done so far has laid the foundation for a climate-resilient agriculture movement in India. The importance of such a movement will be obvious considering the fact that 60 per cent of India's population of 1.1 billion depend upon agriculture for their livelihood. In addition, India has to produce food, feed and fodder for over 1.1 billion human and over a billion farm animal populations.

Challenges ahead

2010-11 is a watershed year in the history of Indian agriculture. Producing food in adequate quantities and making them available at affordable prices will be the greatest challenge during this year. Also, our food security should be built on the foundation of home-grown food, since agriculture is the backbone of the livelihood security system of nearly 700 million people in the country. Nearly 60 per cent of the cultivated area is rain-fed and these are the areas where pulses, oilseeds and other crops of importance to nutrition security, such as millets, are grown. I need hardly emphasise that India is the home for the largest number of malnourished children, women and men in the world. The majority of the malnourished are producer-consumers (i.e., farmer-consumers) and landless labour. Increasing the productivity and profitability of small farms is the most effective method of achieving the UN Millennium Development Goal No. 1: reducing hunger and poverty.

Road map

A road map for our agricultural renaissance and agrarian prosperity was presented by the National Commission on Farmers in five reports presented between 2004 and 2006. The reports are yet to be printed, let alone implemented. For example, 70 per cent of India's population does not find a place in the Padma awards announced on 26 January each year, although NCF had stressed the need for according social prestige and recognition to farmers through such gestures. Farming, particularly in the heartland of the Green Revolution comprising Punjab, Haryana and western Uttar Pradesh, is in deep ecological and economic crises.

Some of the areas needing immediate attention and action are discussed below.

Defending the gains already made in the Green Resolution areas through conservation farming, involving concurrent attention to soil health enhancement, water conservation and effective use, biodiversity protection and launching of a climate-resilient agriculture movement is an urgent task. These are the areas which feed the public distribution system. NCF had recommended the allocation of Rs.1000 crore for this purpose. Expenditure in this area will also come under the Green Box provision of WTO. Climate-resilient agriculture will involve shifting attention to per day rather than per crop productivity.

There is immense untapped production potential in eastern India, the sleeping giant of Indian agriculture — Bihar, Chattisgarh, Jharkhand, eastern Uttar Pradesh, West Bengal, Assam and Orissa. A large number of government schemes with substantial financial outlays, like the Rashtriya Krishi Vikas Yojana, the Food Security Mission, and the National Horticulture Mission exist, but are not making the desired impact on the productivity and production of small farmers. A well-planned movement to bridge the yield gap needs to be initiated with the active involvement of farming families and Gram Sabhas.

The gap between potential and actual yields with the technologies on the shelf ranges from 200 to 300 per cent in these areas. Prime Minister Rajiv Gandhi initiated a dry- land farming revolution in these areas through the Pulses and Oilseeds Missions, but the end-to-end approach he had designed was soon given up and there has been a reversion to the business-as-usual approach. I suggest that during 2010-11, 60,000 Pulses and Oilseed Villages may be organised in rain-fed areas, to mark the 60th anniversary of our Republic. In each of these villages, there should be a lab to land programme organised by the Indian Council of Agricultural Research and State agricultural universities. Such villages can be developed with the help of Gram Sabhas and with the active involvement of farm scientists with the requisite knowledge and experience. NREGA workers can help in water harvesting, watershed management and soil health enhancement. Integrated attention to conservation, cultivation, consumption and commerce should be paid. Assured and remunerative marketing will hold the key to stimulating and sustaining farmers' interest. Today, the consumer is paying very high prices for pulses, but the producer lives in poverty.

Conferring the economy and power of scale to farm families with small holdings is the most serious challenge facing our agriculture. Farm size is declining and 70 per cent of farmers cultivated less than 1 ha in 2003, compared with 56 per cent in 1982. Cooperative farming has been successful in the dairy sector in Gujarat and a few other States. It has not been successful in crop husbandry, although Andhra Pradesh has recently initiated a programme for promoting farm cooperatives. There is increasing feminisation of agriculture with 83 per cent of rural female workers engaged in work related to crop and animal husbandry, fisheries and forestry. Gender specific needs of women farmers, including credit, technology, training and support services like crèches and daycare centres, are urgently needed. It is a matter of satisfaction that the Mahila Kisan Sashakthikaran Pariyojana started by MSSRF in Vidarbha three years ago for the skill and management empowerment of women farmers, including the widows of farmers who had committed suicide, was elevated into a national programme in the Union Budget for 2010-11, with an initial allocation of Rs.100 crore.

Nearly 70 per cent of our population is below the age of 35 and 70 per cent of them live in villages. The future of our agriculture will depend upon attracting and retaining youth in farming. This is one of the principal goals of the National Policy for Farmers (2007). There are several government projects, which if revamped and revitalised, can help to make farming as a profession attractive to educated youth. A new programme for youth in agriculture may be initiated by integrating several ongoing schemes like the Small Farmers' Agri-business Consortium (SFAC), Agri-Clinics, Agri-business Centres, Food Parks, etc, This will help to stimulate the formation of Young Farmers' Self-help Groups. SFAC could be developed into a Young Farmers' Agri-business Consortium, bringing together all relevant programmes.

Food and water security will be the major casualties of a rise in mean temperature, monsoon uncertainty, drought, floods and sea level rise. Some of the steps which could be initiated during 2010-11 are:

- Promote a water conservation pond as well as a biogas plant in every farm, wherever there are farm animals.
- Plant one billion fertiliser trees which can serve as soil carbon banks, enrich soil fertility and enhance farm productivity. Funds for this purpose (Rs. 13,000 crore) are available with the Ministry of Environment and

Forest under the Compensatory Afforestation Fund Management and Planning Authority (CAMPA).

- Establish farmer participatory research and training centres for climate change management in each of the 127 agro-climate zones of the country. Such centres will train at least 1 woman and 1 man in every panchayat as Climate Risk Managers. The centres can be located in either Agricultural and Animal Sciences universities or Krishi Vigyan Kendras or ICAR institutes.

- Build mangrove and non-mangrove bioshields along the coast. These are essential for reducing damage from sea level rise, cyclones and tsunamis. Along with the bioshields, 1000 seawater farming demonstrations can be organised. Seawater is a social resource, as stressed by Mahatma Gandhi when he launched the Salt Satyagraha. Seawater farming will involve the establishment of agri-aqua farms. The farmer participatory demonstrations could be organised along the Indian coast as well as in the Lakshadweep and Andaman group of islands.

During this year, we should begin establishing ultra-modern grain storages at least in 50 locations in the country, each with a storage capacity of a million tonnes of foodgrains (i.e., a 50 million tonne storage grid). Government should remain at the commanding height of the food security system.

2010 is a do or die year for Indian agriculture. If we do not take steps to address the serious ecological, economic and social crises facing our farm families, we will be forced to support foreign farmers, through extensive food imports. This will result in a rise in food inflation, increase the rural-urban and rich-poor divides and allow the era of farmers' suicides to persist. On the other hand, we have a unique opportunity for ensuring food for all by mobilising the power of youth and women farmers and by harnessing the vast untapped yield reservoir existing in most farming systems through synergy between technology and public policy.

Overcoming hidden hunger caused by micronutrient deficiencies like iron, iodine, zinc, vitamin A and vitamin B_{12} can be achieved by growing and consuming appropriate local vegetables and fruits. There is a horticultural remedy for every nutritional malady. *Moringa,* which is a jewel in the horticultural crown, is an example.

Urban and non-farming members of the human family should realise that we live on this planet as the guests of sunlight and green plants, and of the farm women and men who toil in sun and rain, and day and night, to produce food for over 6 billion people, by bringing about synergy between green plants and sunlight. Let us salute the farmers of the world and help them to help in achieving the goal of a hunger free world, the first among the UN Millennium Development Goals.

❑❑❑

Chapter 17

Managing Anticipated Food Crises

The Food and Agriculture Organisation of the United Nations (FAO) has alerted developing countries about possible steep rises in food prices during 2011, if steps are not taken immediately to significantly increase the production of major food crops. According to FAO, 'with the pressure on world prices of most commodities not abating, the international community must remain vigilant against further supply shocks in 2011'. World cereal production is likely to contract by 2 per cent and global cereal stocks may decline sharply. The price of sugar has reached a 30-year high, while international prices of wheat have increased by 12 per cent in the first week of December 2010 as compared to their November average.

The quantitative and qualitative dimensions of under- and malnutrition prevailing in our country are well known. The Steering Committee of a high level panel of experts on food and nutrition set up under my chairmanship to advise the UN Committee on Food Security (CFS) has recently concluded that what we need urgently is a comprehensive, coordinated approach to tackling chronic, hidden and transitory hunger, and not piecemeal approaches. This is also the lesson we can learn from countries which have been successful in combating hunger such as Brazil, which under its "Zero

Hunger" programme has achieved convergence and synergy among numerous nutrition safety net programmes. To some extent, this is what is being attempted under the proposed National Food Security (or Entitlements) Act of the Government of India.

What should be our priority agenda for 2011 on the food front ? At least six areas need urgent and concurrent attention. First, the National Policy for Farmers placed in Parliament in November 2007, on the basis of a draft provided by the National Commission on Farmers (NCF), should not continue to remain a piece of paper, but should be implemented in letter and spirit. This is essential to revive farmers' interest in farming. Without the wholehearted involvement of farmers, particularly of the young as well as women farmers, it will be impossible to implement a Food Entitlements Act in an era of increasing price volatility in the international market. The major emphasis of the National Policy for Farmers is imparting an income orientation to agriculture through both higher productivity per units of land, water and nutrients, and assured and remunerative marketing opportunities. The Green Revolution of the 1960s was the product of interaction among technology, public policy and farmers' enthusiasm. Farmers, particularly in northwest India, converted a small government programme into a mass movement. The goal of food for all can be achieved only if there is similar enthusiastic participation by farm families.

Second, every State government should launch a "bridge the yield gap" movement, to take advantage of the vast untapped yield reservoir existing in most farming systems even with the technologies currently on the shelf. This will call for a careful study of the constraints — technological, economic, environmental and policy — responsible for this gap. The Rs.25,000 crore Rashtriya Krishi Vikas Yojana of the Government of India provides adequate funding for undertaking such work both in irrigated and rain-fed areas. Enhancing factor productivity leading to more income per unit of investment on inputs will be essential for reducing the cost of production and increasing the net income. Scope for increasing the productivity of pulses and oilseed crops is particularly great. The programme for establishing 50,000 Pulses and Oilseed Villages included in the Union Budget for 2010-11 is yet to be implemented properly. The cost of protein in the diet is going up and the Pulses Villages will help to end protein hunger.

There are outstanding varieties of chickpea, pigeon pea, *moong*, *urad* and other pulses available now. What is important is to multiply the good strains and cultivate them with the needed soil health and plant protection measures. The gap between demand and supply in the case of pulses is nearly 4 million tonnes. We should take advantage of the growing interest among farmers in the cultivation of pulses, both due to the prevailing high prices and due to these crops requiring less irrigation water. Such high value, but low water requiring crops also fix nitrogen in the soil. Before the advent of mineral fertilisers, cereal-legume rotation was widely adopted for soil fertility replenishment and build-up.

Third, the prevailing mismatch between production and post-harvest technologies should be ended. Safe storage, marketing and value addition to primary products have to be attended to at the village level. Home Science Colleges can be enabled to set up Training Food Parks for building the capacity of Women Self-Help Groups in food processing. A national grid of ultra-modern grain storage facilities must be created without further delay. In addition to over 250 million tonnes of foodgrains, we will soon be producing over 300 million tonnes of fruits and vegetables. Unless processing and storage are improved, post-harvest losses and food safety concerns will continue to grow.

We should also expand the scope of the Public Distribution System by including in the food basket a whole range of underutilised plants like millets and, where feasible, tubers. NCF pointed out that eastern India is a sleeping giant in the field of food production. The sustainable management of the Ganges Water Machine (this term was first used by the late Professor Roger Revelle) will provide uncommon opportunities for an ever-green revolution in this area. Fortunately Chief Minister Nitish Kumar is taking steps to make Bihar the heartland of the ever-green revolution movement in this region. The Ganges Water Machine is capable of helping us to increase food production considerably, provided we utilise groundwater efficiently during *rabi* and replenish the aquifer during *kharif*.

Fourth, a nutrition dimension should be added to the National Horticulture and Food Security Missions. Hidden hunger caused by the deficiency of micronutrients like iron, iodine, zinc, vitamin A and vitamin B_{12} can be overcome at the village level by taking advantage of horticultural remedies for nutritional maladies. Popularisation of multiple-fortified salt will also be valuable, since this is both effective and economical.

Fifth, a small farm management revolution which will confer the power and economy of scale on farmers operating one hectare or less is an urgent need. There are several ways of achieving this and these have been described in detail in the chapter titled, "Farmers of the 21st Century" in the NCF report. We need to foster the growth of a meaningful services sector in rural India, preferably operated by educated young farmers. The services provided should cover all aspects of production and post-harvest operations. Group credit and group insurance will be needed. Contract farming can be promoted if it is structured on the basis of a win-win situation both for the producer and the purchaser.

Finally, there is need for proactive action to minimise the adverse impact of unfavourable changes in climate and monsoon behaviour and to maximise the benefits of favourable weather conditions. For enabling farmers to develop a "we shall overcome" attitude in the emerging era of climate change, we need to set up in each of the 128 aqro-climatic zones identified by the Indian Council of Agricultural Research a Climate Risk Management Research and Training Centre. These Centres should develop alternative cropping patterns to suit different weather probabilities. They should develop methods of checkmating potential adverse conditions. Along with a climate literacy movement, a woman and a male member of every Panchayat and Nagarpalika will have to be trained as Climate Risk Managers. We will then have over half a million trained Climate Risk Managers, well versed in the science and art of climate change adaptation and mitigation. Such a trained cadre of grass-root Climate Risk Managers will be the largest of its kind in the world.

The present year is ending with damage to rice and other crops in Andhra Pradesh and Tamil Nadu due to excess rain towards the end of the crop season. Farming is the riskiest profession in the world since the fate of the crop is closely linked to the behaviour of the monsoon. Even if there is assured irrigation, natural calamities like cyclones, hailstorms and very heavy showers take their toll. The National Monsoon Mission proposed to be taken up with the participation of US expertise will certainly help to refine the prediction of weather as well as the status of crops and commodity prices. Also, the Mahatma Gandhi National Rural Employment Guarantee Programme provides unique opportunities for strengthening our water security system through scientific rainwater harvesting and watershed management. This valuable benefit can however be realised only by integrating technology with labour. Once a national grid of Climate Risk Management Research and Training Centre comes into existence, it will be possible to build up seed banks of alternative crops, which can be grown if the first crop fails. Drought and Flood Codes should specify the action possible at the end of such calamities. For example, in the flood affected areas, soil moisture will be adequate to grow a short duration fodder crop or a vitamin A-rich sweet potato.

Eternal vigilance is the price of stable agriculture. Early warning helps to take timely action. Food and water security will be the most serious casualties of climate change. 2012 will be a test case to assess whether we as a nation are capable of initiating proactive action to meet the challenges of price volatility, chronic hunger, agrarian despair and climate change.

❑❑❑

Chapter 18

To the Hungry, God is Bread

The National Food Security Bill, 2011 designed to make access to food a legal right, is the last chance to convert Gandhiji's vision of a hunger-free India into reality.

Mahatma Gandhi's articulation of the role of food in a human being's life in his speech at Noakhali, now in Bangladesh, in 1946 is the most powerful expression of the importance of making access to food a basic human right. Gandhiji also wanted that the pathway to ending hunger should be opportunities for everyone to earn their daily bread, since the process of ending hunger should not lead to the erosion of human dignity. Unfortunately, this message was forgotten after the country became independent in 1947, and government departments started referring to those being provided some form of social support as "beneficiaries". The designation "beneficiary" is also being applied to the women and men who toil for 8 hours in sun and rain under the Mahatma Gandhi National Rural Employment Guarantee Programme (MGNREGA). Sixty-five years after Gandhiji's Noakhali speech, we find that India is the home for the largest number of under- and malnourished children, women and men in the world. There are more persons going to bed partially hungry now, than the entire population of India in 1947.

Recent articles of P. Sainath in *The Hindu* (September 26 and 27, 2011) bring out vividly the extent of deprivation and destitution prevalent in rural India. Rural deprivation and agrarian distress lead to the growth of urban slums and suffering. The recent submission of the Union Planning Commission to the Supreme Court on the amount needed per day per person in urban and rural India for meeting his/her needs in the areas of nutrition, education and health care (ie. Rs.35 per person per day in urban India and Rs.26 in rural India) has shown how divorced this important organisation has become from the real life of the poor. It is in this context that there is at least a ray of hope in the draft National Food Security Bill, 2011 placed on the website of the Ministry of Consumer Affairs, Food and Public Distribution, now the under the charge of the humanist, Professor K. V. Thomas. This draft will ultimately go through a Select Committee of Parliament and I hope the final version designed to make access to food a legal right, rather than remain a token of political patronage, will help to erase India's current image as the land of the malnourished. The stated aim of the draft Bill is "to provide for food and nutritional security, in human life- cycle approach, by ensuring access to adequate quantity of quality food at affordable prices, for people to live a life with dignity". To realise this goal, we must ensure that every child, woman and man has physical, economic and social (in terms of gender) access to balanced diet (ie, the needed calories and protein), micronutrients (iron, iodine, zinc, vitamin A, vitamin B_{12}, etc.) as well as clean drinking water, sanitation and primary health care.

A life-cycle approach to food security will imply attention to the nutritional needs of a human being, from conception to cremation. The most vulnerable but most neglected segment is the first 1000 days in a child's life. This covers the period from conception to the first two years in the life of the child. This is the period when much of the brain development takes place. Obviously the child during this period can be reached only through the mother. The life-cycle approach to food security, hence, starts with pregnant women. The high incidence of children with low birth weight (i.e., less than 2.5 kg. at birth) is the result of maternal and foetal undernutrition. Such children suffer from several handicaps in later life, including impaired cognitive ability. Denying a child even at birth an opportunity for the full expression of its innate genetic potential for physical and mental development is the cruelest form of inequity. The Integrated Child Development Services (ICDS) will have to be redesigned and implemented in two time frames (0-2 and 3-6 years).

From the view point of legal rights, the draft Bill addresses only the issue of economic access to food. The other two components of food security, namely, availability of food, which is a function of production, and absorption of food in the body, which is a function of clean drinking water, sanitation and primary health care, cannot easily be made into legal entitlements. To make food for all a legal right, it will be necessary to adopt a Universal Public Distribution System (PDS) with common but differentiated entitlements with reference to the cost and quantity of foodgrains. The draft Bill adopts the nomenclature suggested by the National Advisory Council (NAC) and divides the population into *priority*, i.e., those who need adequate social support, and *general*, i.e., those who can afford to pay a higher price for foodgrains. The initial prices proposed are Rs.3, 2 and 1 per kg for rice, wheat and millet, respectively, for the priority group, and 50 per cent of the minimum support price for the general group. In a Universal PDS system, both self-selection and well- defined exclusion criteria operated by elected local bodies will help to eliminate those who are not in need of social support for their daily bread. In fact, it is the general group that should be supporting financially the provision of highly subsidised food to the economically and socially under-privileged sections of our society. In the case of the well-to-do, the aim of the Universal PDS should be to ensure physical access to food.

The widening of the food basket by including a wide range of nutri-cereals (normally referred to as "coarse cereals"), along with wheat and rice, is a very important feature of the Food Security Bill. Nutri-cereals like *bajra, ragi, jowar,* maize, etc., constitute "health foods" and their inclusion in PDS, along with wheat and rice, will help to increase their production by farmers. Nutri-cereals are usually cultivated in rainfed areas and they are also more climate resilient. Hence, in an era of climate change, they will play an increasingly important role in human nutrition security. During 2010-11, our farm women and men produced 86 million tonnes of wheat, 95 million tonnes of rice and 42 million tonnes of nutri-cereals or coarse cereals. The production of nutri-cereals, grown in dry farming areas, will go up if procurement and consumption go up. Thus, the addition of these grains will help to strengthen concurrently foodgrain availability and nutrition security.

The other components of the Bill, which do not involve legal commitments, refer to agricultural production, procurement and safe storage of grains, clean drinking water and sanitation. The temptation to provide cash instead of grains to the priority group should be avoided. Currency notes can be printed, but grains can be produced only by farmers, who constitute nearly two-thirds of our population. Giving cash will reduce interest in procurement and safe storage. This in turn will affect production. The "Crop Holiday" declared by farmers in the East Godavari District of Andhra Pradesh is a wake-up call. A Committee, chaired by Dr. Mohan Kanda, set up by the Government of Andhra Pradesh has pointed out that the following are some of the factors which formed the basis of the decision of a large number of farm families not to grow rice this *kharif* season. First, the MSP presently offered does not cover the cost of production; the MSP fixed by the Government of India was Rs.1080 per quintal for common varieties, while the cost of production was Rs.1270 per quintal. Secondly, procurement is sluggish since it is largely being done by the rice mills. Third, late release of canal water, non-availability of credit and other essential inputs and delayed settlement of crop insurance dues are also affecting the morale and interest of farm families. Thus farmers are facing serious economic, ecological and farm management difficulties. The government should seriously consider adopting as a general policy the formula suggested by the National Commission on Farmers that MSP should be C2 plus 50 per cent (ie, total cost of production plus 50 per cent).

Finally, the Bill provides for the setting up of Food Security Commissions at the State and Central level. The two essential ingredients of success in implementing the legal right to food are political will and farmers' skill. Hence, it will be appropriate if the State Food Security Commissions are chaired by farmers with an outstanding record of successful farming. They will then help to ensure adequate food supply to feed the PDS. At the national level, the following composition proposed by the National Commission on Farmers (NCF) in their final report submitted in October 2006 would help to ensure adequate political will and oversight. NCF's suggestion was to set up a National Food Security and Sovereignty Board at the central level, with the Prime Minister as Chair. The other members could be the concerned Ministers of the Central Government, leaders of political parties in Parliament, a few Chief Ministers of surplus and deficit States and a few leading farmers and experts. Unless we develop and introduce methods of ensuring effective political and farmers' participation in implementing successfully the Food Security Bill, we will not be able to overcome the problems currently faced by PDS at some places arising from corruption in the distribution of entitlements.

The National Food Security Bill, 2011 provides the last chance for making a frontal attack on poverty-induced hunger and for realising Mahatma Gandhi's desire that the God of Bread should be present in every home and hut in our country. We should not miss this opportunity.

❑❑❑

Chapter 19

The Wheat Mountains of the Punjab

It was in April-May 1968 that the country witnessed the wonderful spectacle of large arrivals of wheat grain in the *mandi*s of the Punjab like Moga and Khanna. Wheat production in the country rose to nearly 17 million tonnes that year, from the previous best harvest of 12 million tonnes. Indira Gandhi released a special stamp titled "Wheat Revolution" in July 1968, to mark this new phase in our agricultural evolution. The nation rejoiced at our coming out of a "ship to mouth" existence. Later in 1968, Dr. William Gaud of the U.S. referred to the quantum jumps in production brought about by semi-dwarf varieties of wheat and rice as a "Green Revolution." This term has since come to symbolise a steep rise in productivity and, thereby, of production of major crops.

Wheat production this year may reach a level of 85 million tonnes, in contrast to the 7 million tonnes our farmers harvested at the time of Independence in 1947. I visited several grain *mandi*s in Moga, Khanna, Khananon and other places in the Punjab in April 2011 and experienced concurrently a feeling of ecstasy and agony. It was heart-warming to see the great work done by our farm men and women under difficult circumstances when, often, they had to irrigate the fields at night due to lack of availability of power during the day. The cause of agony was the way the grains produced by farmers with loving care were being handled. The various State marketing agencies and the Food Corporation of India (FCI) are trying their best to procure and store the mountains of grains arriving every day. The gunny bags containing the wheat procured during April-May 2010 are still occupying a considerable part of the storage space available at several *mandis*. The condition of the grains of earlier years presents a sad sight. The impact of moisture on the quality of paddy is even worse. Malathion sprays and fumigation with Aluminium Sulphide tablets are used to prevent grain spoilage. Safe storage involves attention to both quantity and quality. Grain safety is as important as grain saving. Due to rain and relatively milder temperature, grain arrivals were initially slow, but have now picked up. For all concerned with the procurement, dispatch and storage of wheat grains in the Punjab-Haryana-western Uttar Pradesh region, which is the heartland of the Green Revolution, the task on hand is stupendous.

Farmers in the Punjab contribute nearly 40 per cent of the wheat and 26 per cent of the rice needed to sustain the public distribution system. The legal entitlement to food envisaged under the proposed National Food Security Act cannot be implemented without the help of the farm families of the Punjab, Haryana and other grain-surplus areas. Farmers are facing serious problems during production and post-harvest phases of farming due to inadequate investment in farm machinery and storage infrastructure. The investment made and steps taken to ensure environmentally sustainable production and safe storage and efficient distribution of grains will determine the future of both agriculture in the Punjab and national food security.

On the production side, the ecological foundations essential for sustainable food production are in distress. There is an over-exploitation of the aquifer and nearly 70 per cent of irrigated area shows a negative water balance. The quality of the water is also deteriorating due to the indiscriminate use of pesticides and mineral fertilisers. Over 50,000 ha of cropland in the south-west region of the Punjab are affected by waterlogging and salinisation. Deficiencies of nitrogen, phosphorous and zinc are affecting 66, 48 and 22 per cent of soils in the Punjab, respectively. No wonder factor productivity, i.e., return from a unit of input, is going down. Unless urgent steps are taken to convert the Green Revolution into an ever-green revolution leading to the enhancement of productivity in perpetuity without associated ecological harm, both agriculture in the Punjab and our public distribution system will be in danger. Worried about the future fate of farming as a profession, the younger generation is unwilling to follow in the footsteps of their parents and remain on the farm. This is the greatest worry. If steps are not taken to attract and retain youth in farming, the older generation will have no option but to sell land to real estate agents, who are all the time tempting them with attractive offers. Global prices of wheat, rice and maize are almost 50 per cent higher than the minimum support price paid to our farmers. Our population is now over 1.2 billion and we can implement a sustainable and affordable food security system only with home-grown food.

A disturbing finding of Census 2011 is the deteriorating sex ratio in the Punjab-Haryana region. The female-male ratio among children has come to its lowest point since Independence. Already, women are shouldering a significant portion of farm work. If the current trends of youth migrating from villages coupled with a drop in the sex ratio continue, agricultural progress will be further endangered. The prevailing preference for a male child is in part due to the fear of farmland going out of a family's control, when the girl child gets married. I hope the loss of interest in taking to farming as a profession among male youth will remove the bias in favour of male children. I foresee an increasing feminisation of agriculture in the Green Revolution areas. While the drop in the sex ratio should be halted, steps are also needed to intensify the design, manufacture and distribution of women-friendly farm machinery.

I would like to offer some suggestions on what needs to be done immediately for ensuring sustainable production coupled with efficient procurement, storage and distribution of foodgrains.

The first task is to defend the gains already made in improving the productivity and production of wheat, rice, maize and other crops. For the purpose of providing the needed technologies, it will be advisable to set up soon a Multi-disciplinary Research and Training Centre for Sustainable Agriculture at the Punjab Agricultural University, Ludhiana. This centre can be organised under the National Action Plan for the Management of Climate Change developed under the chairmanship of the Prime Minister, which includes a Mission for Sustainable Agriculture. Such a centre should initiate a Land and Water Care Movement in the Punjab in association with the farming community. The other urgent task is the promotion of appropriate changes in land use. Over 2.7 million ha are now under rice, leading to the unsustainable exploitation of the groundwater. Our immediate aim should be to find alternative land use for about a million ha under rice. This will be possible only if farmers can get income similar to that they are now earning from rice. Possible alternative crops will be maize and *arhar* (pigeon pea). Quality protein maize will fetch a premium price from the poultry industry which is fast growing in the Punjab. *Arhar* being a legume will also enrich soil fertility as well as soil physical properties. Other high value but low water requiring crops like pulses and oilseeds can also be promoted. At the same time, there could be diversified basmati rice production in over a million ha. In addition to Pusa Basmati 1121 which occupies the largest area now, Pusa Basmati I (1460) and Pusa Basmati 6 (1401) can be promoted. These have resistance to bacterial leaf blight. Varietal diversity will reduce genetic vulnerability to pests and diseases.

For handling the over 26 million tonnes of wheat which will be purchased during this season, a four-pronged strategy may be useful. First, distribution through railway wagons could be expanded and expedited. One wagon can handle 2,500 tonnes. Currently 30,000 to 40,000 tonnes of wheat are being dispatched each day through wagons. With advanced planning, this quantity can be raised to over 1 lakh tonnes per day. They can be dispatched to different States for meeting the needs of PDS, ICDS, School Noon-Meal Programme, Annapoorna, etc. Second, the present Common Agricultural Policy (CAP) and godown storage systems can be improved with a little more investment and planning. In the Punjab there are 146 *mandis* and 1,746 purchase points. They could be grouped and their infrastructure improved. Third, storage in modern silos, like the one put up at Moga by Adani Agri-logistics, and another one coming up in Amritsar, should be promoted. This will help to adopt an end-to-end system from the point of view of procurement, cleaning, quality assurance, safe storage and distribution. The cost of building silos to store a million tonnes of foodgrains may be about Rs.600 crore, if the required land is made available by State governments. An investment of about Rs.10,000 crore would help to establish a grid of modern grain storages with a capacity for storing, in good condition, over 15 million tonnes in the Punjab-Haryana-western U.P. region. Lastly, export options can be explored after taking steps to make food available to the hungry, as suggested by the Supreme Court. Also, we should ensure that adequate foodgrains will be available for implementing the proposed Food Security Act. Export should be done only if the global food prices are attractive and if the profit made is distributed as bonus to our farmers, as suggested by the National Commission on Farmers.

It is time that we organise a national grid of grain storages, starting with storage at the farm level in well-designed bins and extending to rural godowns and regional ultra-modern silos. Post-harvest losses can then be minimised or even eliminated, and food safety ensured. Unless the prevailing mismatch between production and post-harvest technologies is ended, neither the producer nor the consumer will derive full benefit from bumper harvests.

❑❑❑

Chapter 20

Fish for All and For Ever

Three years after the establishment of the Central Institute of Fisheries Education (CIFE), a global conference was held at Rio de Janeiro, Brazil, for promoting environmentally sustainable development in all spheres of human activity. At this UN Conference on Environment and Development, generally referred to as the Earth Summit, an Agenda 21 was adopted to provide guidelines for mainstreaming ecology in all major research and development programmes. In Agenda 21, pride of place has been given to the management of aquatic resources from both the quantitative and qualitative aspects. 2012 marks the twentieth anniversary of the Rio conference. It will be useful to prepare a balance sheet indicating where we have succeeded and where we have failed in implementing Agenda 21 in the area of inland and marine fisheries, as well as integrated coastal zone management.

Agenda 21

Articles 17 and 18 of the Rio Agenda 21 relate to

- Protection of the oceans and coastal areas and the conservation and sustainable use of their living resources (Article 17).

- Protection of the quality and supply of freshwater resources; application of integrated approaches to the development, management and use of water resources (Article 18).

Both these action points are important for sustainable fisheries. We should also harness modern technologies for helping our fisher communities. For example, mobile telephony is a transformational technology with reference to small-scale fisheries. The Indian National Centre for Ocean Information Services (INCOIS) provides data on wave heights from different distances from the shoreline. Data are also available on the location of fish shoals. The M. S. Swaminathan Research Foundation started, in partnership with Qualcomm, Tata Teleservices, Astute Technology System and Village Panchayats and Fishing Communities, a Fisher Friend Project in 2007 to take relevant information to the fisher communities before they enter the sea in their small boats. MSSRF created a system of information flows whereby fishermen can gain access to fishing-related information such as wave heights, weather forecasts, high potential fishing zones and market prices. This service has helped to revolutionise the lives and livelihoods of fisher communities. Induced breeding is another area where Indian fisheries scientists have made significant contributions.

CIFE should develop more such transformational technologies. For this purpose, CIFE should also initiate inquiry-based instruction. The journal *Science* has instituted a new prize to recognise contributions in the field of inquiry-based teaching. According to Dr Bruce Alberts, Editor-in-Chief, *Science*: 'The new award has been stimulated by the fact that the world badly needs a revolution in science education — a revolution that must begin at the college level.' CIFE can organise classes for fisher communities on the objectives of the Coastal Regulation Zone (CRZ) Notification, 2011, which replaces the 1991 notification. Unlike the 1991 notification, the 2011 regulation takes into account both the seaward and landward sides of the coast. This will help to keep the coastal waters free of pollution. The disposal of wastes and effluents into coastal water is a prohibited activity. Also for the first time, a separate draft Island Protection Zone Notification has been issued in the case of Andaman and Nicobar and the Lakshadweep group of islands. The provision under CRZ-1 has been strengthened to cover all ecologically sensitive areas, such as mangrove wetlands, corals and coral reefs, national parks, etc.

Seawater as a social resource

2010 marked the 80th anniversary of the Salt Satyagraha launched by Mahatma Gandhi in 1930 at Dandi in Gujarat. Nearly, 97 per cent of the world's water is seawater. By protesting against the imposition of a tax on salt manufacture, Gandhiji emphasised that seawater is a social resource. Water will be one of the key limiting factors in food production in the coming decades. Hence, on the occasion of the 80th anniversary of the Dandi March, as well as a similar one organised by C. Rajagopalachari at Vedaranyam in Tamil Nadu, MSSRF organised a Seawater Farming for Coastal Area Prosperity programme. The technology involves agro-forestry systems involving integrated tree-fish farming. Besides mangroves, other salt tolerant trees and shrubs like *Salicornia, Atriplex, Sesuviun, Casuarina,* etc., can be cultivated along with fish ponds. In the Kuttanad area of Kerala, farmers have been cultivating rice at 2.6 m below mean sea level for over a century. A rice-fish (either carps or giant prawns) farming system is now becoming popular in this area. Thus, both seawater farming and below sea level farming are feasible. The unique below sea level farming system of Kuttanad is being proposed for recognition under FAO's Globally Important Agricultural Heritage sites. Also, a genetic garden of halophytes is being developed. Halophytes offer opportunities for converting seawater into a valuable source for producing products of agricultural and human nutrition value.

A Fish for All Research and Training Centre has been established at Kaveripoompattinam in the Nagapattinam district of Tamil Nadu. This area was affected by the tsunami in December 2004. At this centre, training is imparted in all aspects of fish capture, culture, processing and marketing. Food safety aspects receive particular attention. Such capacity-building centres based on concurrent attention to all links in the capture / culture to consumption chain are needed all over the country.

The 2010 collapse of the Bolivian fisheries is a wake-up call relating to the potential impact of climate change on marine fisheries. Climate change led to the mass death of fish in rivers and in the Antarctic Ocean. An estimated 6 million fish and thousands of alligators, turtles and river dolphins perished due to a sudden change in water temperature. The water temperature in Bolivian rivers is normally about 15°C. But last July the temperature fell to 4°C. It is not unlikely that the extreme weather conditions in July might have been related to El Nino-Southern Oscillation, as has been pointed out in the September 2010 issue of *Nature.*

We can expect similar catrostophies more frequently in the future since extreme and unpredictable weather conditions are now becoming common. CIFE scientists should develop an anticipatory research programme to checkmate the multiple adverse impact of unfavourable weather on capture and culture fisheries.

Land is a shrinking resource for food production. Fortunately, we have vast oceans as well as rich inland water resources. This will help us to foster a "fish for all and for ever" movement. Conservation and sustainable and equitable use of fish genetic resources will be important for promoting climate-resilient fisheries. Low external input sustainable aquaculture will be important to avoid groundwater pollution in both coastal and inland areas. Recombinant DNA technology offers opportunities for generating novel genetic combinations for facing the challenge of climate change and global warming. However, we need a large number of women and men trained in the biosafety and biosecurity aspects of genetic modification. We need an autonomous, professionally led National Biotechnology Regulatory Authority which inspires confidence that the precautionary principle will be the bottom line of regulatory policies.

Lessons from the recent triple tragedy in Japan

Nearly 25 per cent of our populations live near coastal areas. The devastating earthquake and tsunami which affected parts of Japan in March 2011 underline the importance of giving greater attention to enlarging the coping capacity of coastal communities to natural calamities. The Japanese are used to earthquakes and tsunamis. What was new this time was the impact of the tsunami on several nuclear power plants located in the Fukushima-Daiichi area. The damage done to the nuclear power reactors emphasises the fact that the power of nature could trump technology. We should give thought to methods of enhancing the coping capacity of coastal communities to such mega-disasters as well as develop a plan of action for promoting sustainable capture and culture fisheries both along the coast and in inland waters.

There are three important lessons we can learn from the Japanese experience. First, we should develop and introduce educational tools which can promote high synergy societies. The Japanese system of early childhood education instills the habit of caring for others and helps to foster high social synergy. This strengthens the coping capacity of the people in times of disasters as well as their ability to convert a calamity into an opportunity for greater progress. CIFE should organise non-degree training programmes for fisher communities on methods of group cooperation and promote high synergy fishing communities.

A second lesson we should learn from the Japanese experience is that it is only through harmony with nature that we can minimise the damage arising from natural disasters. There is now public concern about the safety of nuclear power plants located along the coast such as Kalpakkam and Kudankulam in Tamil Nadu. As I have repeatedly said, in addition to appropriate steps in reactor design and engineering, we should promote bio-shields comprising mangrove and non-mangrove species in the coastal areas adjoining nuclear power plants.

Third, while earthquakes and tsunamis are no strangers to the Japanese people, the new threat which has created the greatest fear is the damage done to nuclear power plants. Ever since nuclear power was harnessed after World War II, there have been three major accidents:

- UK's Windscale fire of 1957
- Partial meltdown at Three Mile Island, Pennsylvania, USA, 1979
- Radioactive plume drifting across Europe after the explosion at Chernobyl in Ukraine, 1986

The Fukushima-Daiichi disaster of 2011, which is the fourth most serious nuclear accident, has lead to a serious discussion on the establishment of nuclear power plants in seismically active areas. The global setback for nuclear energy that is likely to follow the Fukushima tragedy will probably encourage greater investment in alternative low carbon and non-carbon energy sources. However, nuclear power is environmentally benign and hence we should continue to explore fail-safe methods of nuclear power generation. Also, the Atomic Energy Regulatory Board (AERB) should be an autonomous one and should not be controlled by those whom the board has to regulate. Our Prime Minister has recently announced that AERB will be given complete autonomy and that it will be free from the control of the Atomic Energy Commission. It is only such autonomous regulatory mechanisms that can inspire public, political, professional and media confidence.

❑❑❑

Chapter 21

Leveraging Agriculture for Improving Nutrition and Health

As it is well understood, food and drinking water are the first among the hierarchical needs of a human being. Growing population, expanding ecological footprint, diminishing per capita land and water availability, increasing biotic and abiotic stresses, and, above all, the prospects for adverse changes in temperature, precipitation and sea level as a result of climate change, emphasise the need for keeping issues relating to agriculture high on the professional, political and public agenda. The multiple roles of agriculture in food production, improving nutrition and health, and climate change mitigation are now well recognised scientifically, but are yet be integrated into coherent national policies and strategies. Opportunities for generating synergy among agriculture, nutrition and health are great and this conference is therefore a timely one.

In India, the relationship between diet and health have been recognised since ancient times in indigenous medical systems like ayurveda. The National Institute of Nutrition of India is affiliated to the Indian Council of Medical Research (ICMR) to ensure synergy between nutrition and health care. Similarly, India was one of the early countries to develop an Integrated Child Development Service (ICDS) involving concurrent attention to nutrition, health and education. In spite of such early recognition of the need to "deliver as one" in relation to the nutritional and health requirements of the population, India has an unenviable record in overcoming child and adult malnutrition and in linking synergistically agriculture, nutrition and health.

Let me cite a few examples of the immense benefits that will accrue from leveraging agriculture for improving nutrition and health. When I was at the International Rice Research Institute, the Philippines, I organised in 1986 a consultation jointly with the World Health Organisation (WHO) on how to avoid the breeding of malarial mosquito in rice fields. We concluded that alternate wetting and drying of rice fields will disrupt the breeding cycle of the mosquito. Such a practice does not affect yield but confers the additional benefit of a substantial reduction in the demand for irrigation water. This approach to water management in rice fields is now incorporated in what is popularly known as the System of Rice Intensification (SRI) — an agronomic management method being popularised by Dr. Norman Uphof of the Cornell University, USA. As President of the Pugwash Conferences on Science and World Affairs, I had organised discussions on the role of nutrition in the treatment of HIV / AIDS patients both in the "first wave" countries like South Africa, and "second wave" countries like India. The experience in both the first and second wave countries was the same — namely, a food-cum-drug approach yields the best results. The same is true with reference to tuberculosis and leprosy where the need for prolonged treatment limits opportunities to poor patients for earning their daily bread.

Several steps are urgently needed for achieving the goal of linking agriculture with nutrition and health. First, nutritional considerations must be incorporated in farming systems research. For example, pulses or grain legumes should find a place in the crop rotations. Crop-livestock integrated production systems as well as coastal and inland capture and culture fisheries will help immensely in ensuring that the needed macro- and micro-nutrients are available in the diet. ICAR's All India Coordinated Project on Farming Systems Research should have a competent nutritionist on its staff, so that appropriate agricultural remedies are introduced for the nutritional maladies of the area.

The National Horticulture Mission affords uncommon opportunities for addressing the problem of micro-nutrient malnutrition, i.e., the deficiency of iron, iodine, zinc, vitamin A, vitamin B_{12}, etc., in the diet. What is important is the addition of the nutritional dimension in the programmes designed to promote the cultivation of vegetables and fruits in different parts of the country. A Home Science graduate well versed in nutrition can be added to the staff of the Mission in every district to promote nutritional literacy in the area.

India proposes to make access to food a legal right soon, through a National Food Security Act. The National Advisory Council has recommended that this Act should have both mandatory rights and enabling provisions. The mandatory rights will include the provision of 35 kg of rice, wheat or millet — pearl millet, sorghum, maize, *ragi* and minor millets (preferred particularly in tribal areas) — per month per family at a price of Rs.1, 2 and 3 per kg in the case of millet, wheat and rice, respectively. The inclusion of nutritious millets, inappropriately called "coarse cereals", will help to improve both nutrition and climate resilience, since these crops are more drought tolerant. The legal entitlements will be structured on a lifecycle basis, with particular attention to the first 1000 days of a child's life (i.e., from conception up to the end of two years). The lifecycle approach will ensure attention to all stages in one's life. ICDS will be restructured so that the nutritional needs of the infant during the first 1000 days are met.

Among the enabling provisions, concurrent attention to clean drinking water, sanitation, environmental hygiene and primary health care, will be an important one. The addition of nutrition in the National Rural Health Mission will help to foster symbiotic linkages among agriculture, nutrition and health. Synergy between nutrition and agriculture will include steps like the cultivation and consumption of *moringa* (drumstick) along with millet. *Moringa* is a nutritional marvel and the millet-cum-*moringa* combination in the diet will help to meet the needs of both macro- and micro-nutrients.

Globally and nationally, the prevailing rates of hunger and malnutrition are inexcusable. There are simple and cost effective approaches to making such a sad situation a problem of the past. This will however require coordinated planning and action among those involved in the agriculture, health and nutrition sectors.

❑❑❑

Chapter 22

GM Food Crops: Risks and Benefits

Transgenic varieties combine genes from totally unrelated species. For example, we can now transfer genes for salinity tolerance to rice from mangroves. The recombinant DNA technology is part of the evolution of genetics starting with the rediscovery of Mendel's Laws of Inheritance in 1900. In the early part of the twentieth century, various techniques like irradiation, use of chemical mutagens and doubling of chromosomes through colchicine treatment were adopted to develop novel genetic combinations.

Today such gene transfer can be done with both precision and ease through recombinant DNA technology. Both molecular marker-assisted breeding and gene transfer now play a very important role in developing genetic combinations to meet the challenges rising from biotic (i.e., pest and diseases) and abiotic (i.e., drought, flood, sea level rise, etc.) stresses. They will gain further importance in the emerging era of climate change. Varieties developed by marker-assisted selection are also eligible for certification under the protocol for products produced through organic farming.

Transgenic varieties may not pose a threat to biodiversity, since the seeds can be kept by farmers. The threat comes from hybrids whose seeds will have to be purchased every year by the farmer. Replacement of numerous local varieties with one or two hybrids or GM (genetically modified) varieties will undermine the sustainability of production, since genetic homogeneity enhances genetic vulnerability to pests and diseases as well as to abiotic stresses.

Transgenic food crops can cause harm to human health if they are not tested very carefully for biosafety aspects. In USA, three different official agencies — FDA (Food and Drug Administration), EPA (Environmental Protection Agency) and APHIS (Agricultural Plant Health Inspection Service) — subject transgenic crops to thorough examination for their potential adverse impact on human health, biodiversity and the environment. It is only after such methodical studies in government laboratories that clearance is given for large-scale cultivation. The US public feels satisfied that the GM crops approved will do no harm. India lacks such facilities at present and public apprehension about the biosafety aspects of GM crops is widespread. ICAR is yet to establish a well-planned all-India coordinated research project for testing GM varieties under isolation, with a national coordinator who is an internationally recognised biosafety expert. We lack the institutional infrastructure necessary for measuring risks and benefits in a transparent manner.

It is popularly said that the Green Revolution was the product of public sector enterprise, while the gene revolution is the result of private sector investment. The first is the result of public good research, while the second is the result of commercial profit research covered by intellectual property rights. What is therefore important is to step up public good research in the field of biotechnology by supporting universities and government research institutions. Fortunately, there is a considerable amount of work in progress in public good research institutions in our country for mainstreaming the principle of inclusivity in relation to access to new technologies. Much work is being done with support from the Department of Biotechnology of the Government of India. Other public good institutions like ICAR, CSIR and ICMR are also undertaking and supporting useful research in harnessing the power of recombinant DNA technology for addressing the issues of resource-poor farmers and consumers. We need to considerably step up such work and give them a pro-nature, pro-poor and pro-women orientation.

Unfortunately, our official mechanisms for examining issues relating to biosafety are inadequate since they do not have their own testing facilities. There is a move to establish a National Biotechnology Regulatory System through an Act of Parliament. Such a regulatory system should be based on a hub-and-spokes model, with the autonomous and professionally-led National Biotechnology Authority serving as the hub and the national and State agencies connected with health, environment, agriculture, etc. serving as spokes.

The National Biotechnology Regulatory System should be capable of thoroughly examining the different aspects of biosafety and biosecurity. First, issues relating to the environment, including the impact on biodiversity, will have to be studied. Second, issues relating to risks and benefits in terms of economics will have to be studied in a transparent and trustworthy manner. Finally, the genetically-modified plants should be subjected to evaluation from the point of view of their chronic effects. In the case of plants like brinjal whose native home is India, every effort should be made to collect and conserve the native germplasm. All efforts in the area of introduction of new technologies should be based on the "4C" principle, namely, conservation, cultivation, consumption and commerce.

Bt brinjal need not be banned, but there should be caution that one or two hybrids do not replace hundreds of native varieties all of which have distinct quality characters. India is the home of brinjal and we have rich genetic diversity in this crop. Steps should be taken to conserve this wonderful gene pool. Also, studies should be carried out on the chronic effects of consuming Bt brinjal throughout one's life. In matters of biosafety and human health, the precautionary principle should be always followed.

Priority should go to solving those problems which cannot be solved with the currently available Mendelian technologies and marker-assisted selection. For example, we need more climate-resilient varieties, such as wheat varieties tolerant to high night temperature, salinity- and drought-resistant plants, and resistance to new pests and diseases. Hybrids are those which exhibit hybrid vigour through a combination of two very different parents. They will however not breed true if grown again. Hybrids could be either conventional or transgenic. In crops like maize and rice, hybrids are used extensively in view of the possibility of producing seeds economically.

Since more than 80 per cent of our farmers have only small holdings, the development of transgenic varieties rather than hybrids should be promoted. Farmers can keep their own seeds of transgenic varieties, as permitted under the National Plant Variety Protection and Farmers' Rights Act.

The basic determinants of successful transgenic plant breeding will be economics, ecology, equity, ethics and employment. In the field of economics, the cost-risk and return structure will determine farmers' choice of technologies investment decisions. In the case of equity, it is important to ensure that all farmers, irrespective of the size of their holding and their innate risk-taking capacity, are enabled to derive advantage from new technologies. In the area of ethics, steps should be taken to adopt the precautionary principle in relation to biosafety assessment. In the area of employment, it will be useful if Womens' Self-Help Groups are formed in villages to produce hybrid seeds. They can be given training in seed technology in Krishi Vigyan Kendras or other institutions where the pedagogic methodology is learning by doing.

A genetic literacy movement should be launched in villages so that farm families are able to fully understand the procedures involved both in Mendelian and molecular breeding. The Department of Biotechnology has initiated a programme for organising DNA clubs in schools. This will have to be expanded to cover adults in rural and tribal areas. Both mistrust and undue expectations arise from inadequate understanding of GM technology and this emphasises the need for a genetic literacy movement among both farmers and urban consumers.

We should not worship or discard a research tool because it is either old or new, but should choose an appropriate mix of Mendelian and molecular approaches to genetic recombination, which can take us to the desired goal surely and safely.

My personal approach to using the science of biotechnology for enhancing human food and nutrition security is what I had proposed in my 2004 report on agricultural biotechnology:

> The bottom line of a National Agricultural Biotechnology Policy should be the economic well-being of farm families, food security of the nation, health security of the consumer, protection of the environment, biosecurity of the country and the security of our national and international trade in farm commodities.

❏❏❏

Chapter 23

Land Rush and Food Security in an Era of Climate Change

On the basis of a proposal I had made three years ago, the Food and Agriculture Organisation of the UN (FAO) launched a Global Soil Partnership for Food Security and Climate Change Adaptation and Mitigation at a Multi-stakeholder Conference held in Rome in September 2011. Even with all the advances made in capture and culture fisheries, nearly 90 per cent of human food requirements will have to come from the soil. Land is becoming a diminishing resource for agriculture, inspite of the growing understanding that the future of food security will depend upon the sustainable management of land resources as well as the conservation of prime farm land for agriculture. The National Commission on Farmers emphasised in its report submitted in 2006 the urgent need for replacing the 1894 Land Acquisition Law with a twenty-first century legislation that safeguards the interests of farmers and farming. Jairam Ramesh is to be complemented for introducing in Parliament a National Land Acquisition and Rehabilitation and Resettlement Bill which pays attention not only to acquisition, but also to the rehabilitation and resettlement of the affected families.

* Member of Parliament (Rajya Sabha) and Chairman, M S Swaminathan Research Foundation

A high level external committee (HLEC) set up under my chairmanship in 2010 by the UN Committee on Food Security (CFS) has recently submitted to CFS a Report on Land Tenure and International Investments in Agriculture. The Report analyses the potential impact of land acquisitions, particularly in Africa, on food security. It has been estimated that about 50 to 80 million hectares of farm land in developing countries have been the subject of negotiations by international investors in recent years, two-thirds of them in sub-saharan Africa, widely recognised as a "hot spot" for endemic hunger. We found little evidence that such large-scale land acquisitions have helped to provide food and jobs to the local population. More than three-quarters of the land deals are yet to demonstrate improvements in agricultural output. The HLEC identified several steps that governments should take towards more effective and equitable land tenure systems, starting with creating more transparent systems for registering, tracking and protecting land rights, in particular of women, tribal families and other vulnerable groups who depend on common property resources for the security of their livelihoods. The satellite and aerial imagery used in biophysical surveys are blind to the rights and institutions that govern how land is actually used on the ground. According to the World Bank the "land rush" is not likely to slow in the future. As a result, the landless labour population will grow leading to greater social unrest in the rural areas of developing countries.

The loss of land for food security has to be measured not only in quantitative terms, but also in respect of land use. According to the US Department of Agriculture, American farmers for the first time will harvest during 2011 more maize for ethanol production than for food or feed. In Europe, about 50 per cent of the rapeseed crop is likely to be used for biofuel production. The plant-animal-man food chain (particularly beef and poultry products) will need several times more land for producing a calorie of meat, as compared to a calorie of cereal or vegetable.

The sudden escalation in the price of rice and wheat observed in 2008 was largely due to a steep increase in the price of fossil fuels leading to a rise in input costs. The growing diversion of farm land for fuel production in industrialised countries, increasing consumption of meat on the part of the affluent sections of the society, and loss of farm land for other uses such as roads, houses and industries are likely to lead to acute food scarcity, severe price volatility and high food inflation by the end of this decade. Several experts have pointed out that "the Arab Spring" had its genesis in high food inflation. This is why I have been stressing that the future belongs to nations with grains and not guns.

On the basis of widespread consultations, FAO has recently prepared voluntary guidelines on the responsible governance of tenure of land, fisheries and forests in the context of national food security. These voluntary guidelines will be considered at the next meeting of CFS scheduled to be held in October 2011. There are elements in these guidelines which are worthy of consideration by the Committee of Parliament, which will go into the provisions of our National Land Acquisition Bill. For example, one provision states, "subject to their national law and legislations and in accordance with national context, states should expropriate only where rights to land (including associated buildings and other structures), fisheries or forests are required for a public purpose. In no way, should expropriation or forced eviction be made for private purposes". The voluntary guidelines also recommend that "states should ensure that women and girls have equal tenure rights and access to land, fisheries and forests, independent of their civil or marital status". Business models should involve steps which will help to generate employment opportunities and strengthen the livelihood security of the poor. "Food security first" should be the motto of our Land Acquisition Bill. Large-scale investment for biofuels is a risk and must be avoided, unless there are instances, as for example in Brazil, where such investments provide a win-win situation for both food and energy security. Land tenure is key to protect land rights. Central and State Governments should have accessible systems for registering, tracking and protecting land rights, including customary rights and common property resources.

In 1981, Member-States of FAO adopted a World Soil Charter, containing a set of principles for the optimum use of world's land resources and for the improvement of their productivity as well as their conservation for future generations. The World Soil Charter called for a commitment on the part of governments and land users to manage the land for long-term advantage rather than a short-term expediency.

International interest in the conservation and management of soil resources for food security and climate change adaptation and mitigation has grown in recent years, because of increasing diversion of farm land for non-farm uses. In May 2011 a Global Soil Forum was formed at a conference held at the Institute for Advanced Sustainability Studies, Potsdam, Germany for enhancing investment on soil resources assessment and management. The Global Soil Forum, with financial support from Germany and a few other donors, will help to identify some key technological options to enhance and sustain soil-based ecosystem services, in order to safeguard food security in the long term. To emphasise the need to conserve soil biodiversity, the European Union has prepared a comprehensive European Soil Biodiversity Atlas.

Over 15 years ago, a Global Water Partnership (GWP) was formed to stimulate attention and action at the national, regional and global levels in the area of sustainable water security. GWP was conferred the status of an international organisation by the Government of Sweden in 2002. India is also a partner in the activities of GWP. Land use decisions are also water use decisions and hence the organisation of a Global Soil Partnership (GSP) to work closely with GWP is a timely initiative. GSP will specifically address urgent problems such as soil degradation, conservation of soil biodiversity, gender and social equity, climate change and soil health management for an evergreen revolution in agriculture. GSP will provide a multi-disciplinary and multi-institutional platform for mobilising the power of partnership in managing threats to food security arising from climate change and "land rush".

Soil anemia also breeds human anemia. Deficiency of micronutrients in the soil results in micronutrient malnutrition in children, women and men, since the crops grown on such soils tend to be deficient in the nutrients needed to fight hidden hunger. With the addition of GSP to the already existing GWP, and with the likely adoption of the Voluntary Guidelines on the Responsible Governance of Tenure of Land and other Natural Resources, we have the global instruments which can assist nations to safeguard and strengthen the ecological foundations of sustainable agriculture and for overcoming endemic, hidden and transient hunger. What is needed is the conversion of global instruments and guidelines into socially sustainable and equitable national regulations, on the lines recommended in the HLEC report on Land Tenure. The National Land Acquisition and Rehabilitation and Resettlement Bill, now under the consideration of Parliament, has a much wider significance than just preventing land grab. The critical role soil plays in food security and climate change adaptation and mitigation has to be widely understood.

Along with oceans, soils offer opportunities for storing carbon. For example, it is estimated that global net primary productivity (NPP) may be about 120 Gt/c/year. Most of it is returned to the atmosphere through plant and soil respiration. If 10 per cent of NPP can be retained in the terrestrial biosphere like wetlands and mangrove ecosystems, 12 Gt/c/year can become a part of a terrestrial carbon bank. Increasing soil C pool by 1 t/c/ha/year in the root zone can increase food production by 30 to 50 million tonnes. Thus, soil carbon banks represent a win-win situation for both food security and climate change mitigation.

Managing our soil and water resources in a sustainable and equitable manner needs a new political vision, which can be expressed through the proposed Land Acquisition Bill. 2012 marks the 20th anniversary of the Rio Earth Summit and 40th anniversary of the Stockholm Conference on the Human Environment. This will be an appropriate occasion to launch a Soil and Water Security Movement through education, social mobilisation through Gram Sabhas, and legislation like the Land Acquisition Bill.

❑❑❑

Chapter 24

Darwinism in a Warming Planet

In 2009, the world celebrated the 200th anniversary of the life and work of Charles Darwin, a transformational scientist who brought about a revolution in our understanding of evolution. Evolution through natural selection among living organisms leading to the survival of the fittest was a concept which was alien to Christian thought in Darwin's time, as faith was strong in the divine creation of the living forms of our planet. Evolution of higher forms of life, including human beings, from fish and other forms of living organisms was an accepted view in ancient Indian thought, as exemplified by the ten forms of manifestation on earth of the Hindu god Vishnu.

"Variety is the spice of life" is a common saying. Variation is a must for selection to occur. Today, we are confronted with the prospect of human-induced changes in climate, leading to adverse variations in temperature, precipitation, flood and sea level. We will have to be prepared for facing the consequences of drought, flood and coastal storms more frequently. Selection of genes for a warming planet has therefore become an urgent task. Fortunately, there is considerable variability in nature with reference to adaptation to new climatic conditions. Thus, halophytes, which are resistant to salinity, and xerophytes, which are resistant to moisture stress, occur in nature. This is why Mahatma Gandhi said that 'nature provides for everyone's need, but not for everybody's greed.'

Gregor Mendel, who propounded the laws of inheritance or genetics, published his work in 1865, six years after the publication of Darwin's *On the Origin of Species.* Mendelian genetics, now reinforced by molecular genetics, helps us to create new genetic combinations capable of surviving under the adverse circumstances created by global warming. Molecular genetics, which helps us to move genes across sexual barriers, has validated the truth behind the statement of Charaka, the time-honoured father of ayurveda, that there is no useless plant or animal in the world. Thus, scientists at MSSRF have been able to develop salt-tolerant varieties of rice by transferring genes from the mangrove species. *Avicennia marina*, and drought- tolerant varieties of rice, using genes from *Prosopis juliflora.* Such novel genetic combinations help us to take advantage of Darwin's concept of the survival of the fittest.

Halophytic plants and salinity-tolerant crops like the genetically modified rice of MSSRF can help to launch a seawater farming movement along our vast coastline. Agro-forestry systems involving the cultivation of mangroves, *Salicornia, Sesuvium, Atriplex* and other salt-tolerant shrubs and trees together with mariculture involving the cultivation of fish will open up new livelihood opportunities to coastal communities. Salt-cum-flood-tolerant rice varieties can also be developed by incorporating the mangrove gene into floating or flood-tolerant varieties of rice.

Modern information technology has opened up a new chapter in artisanal fisheries. Artisanal or small-scale fisheries can become economically and technologically attractive by using cell phones to disseminate data on wave heights and the location of fish shoals. Also, the seawater farming methodology can help to increase yield and income from coastal aquaculture.

By linking the Darwinian concept of evolution with the principles and tools of Mendelian and molecular genetics, we can not only safeguard our food security in an era of climate change, but also strengthen the ecological security of coastal areas and the livelihood security of coastal communities.

The food inflation prevailing in the country is partly due to the high cost of pulses. These protein-rich crops are grown in rain-fed areas, which constitute 60 per cent of our cultivated area. Available data indicate that we can double the yield of pulse crops like *arhar, moong, urad, chenna,* etc., by introducing an integrated package involving attention to soil health enhancement, water harvesting and efficient water use, use of improved seeds and agronomic management, credit and insurance and, above all, assured and remunerative marketing. These components of the Pulses Revolution strategy should be incorporated in the 60,000 Pulses and Oilseed Villages included by the Finance Minister in the 2010-11 Budget. If this programme is implemented properly with the active participation of farm families, we can easily produce the additional 4 million tonnes of pulses we urgently need. Let us convert the calamity associated with climate change into an opportunity for spreading conservation and climate-resilient farming methods.

❑❑❑

Chapter 25

Climate and Food Security

The principle of common but differentiated responsibilities is the core of the many climate agreements arrived at so far, including the Kyoto Protocol (1997) and the Bali Plan of Action (2007). The differentiated responsibilities aim to meet the special needs of developing countries for accelerated and equitable economic development. Both at Copenhagen and Cancun, the industrialised countries proposed limiting the rise in mean temperature to 2°C above normal. Even this seems to be unattainable in the context of the present rate of emission of greenhouse gases (GHG). Hence, the principle of common but differentiated impact of the 2° change in mean temperature is essential for prioritising climate victims. For example, small islands like Tuvalu in the Pacific Ocean, Maldives, Lakshadweep, and Andaman and Nicobar, as well as the Sunderbans in West Bengal, Kuttanad in Kerala and many other coastal locations will all face the prospect of submergence. Floods will become more serious and frequent in the Indo-Gangetic plains. Drought-induced food and water scarcity will become more acute. South Asia, sub-Saharan Africa and the small islands will be the worst victims. In contrast, countries in the northern latitudes will benefit due to longer growing seasons and higher yields.

Addressing the World Climate Conference held in Geneva in 1989 on the theme, "Climate Change and Agriculture", I pointed out the serious implications of a rise of 1 to 2°C in mean temperature on crop productivity in South Asia and sub-Saharan Africa. An expert team constituted by FAO in its report submitted in September 2009, also concluded that for each 1°C rise in mean temperature, wheat yield losses in India are likely to be around 6 million tonnes per year, or around $ 1.5 billion at current prices. There will be similar losses in other crops and our impoverished farmers could lose the equivalent of over US $ 20 billion in income each year. Rural women will suffer more since they look after animals, fodder, feed and water.

We are in the midst of a steep rise in the price of essential food items like pulses, vegetables and milk. The gap between demand and supply is high in pulses, oilseeds, sugar and several vegetable crops including onion, tomato and potato. Production and market intelligence as well as a demand-supply balance based on an integrated import and export policy are lacking. The absence of a farmer-centric market system aggravates both food inflation and rural poverty. FAO estimates that a primary cause for the increase in the number of hungry persons, now exceeding over a billion, is the high cost of basic staples. India has unfortunately the unenviable reputation of being the home for the largest number of undernourished children, women and men in the world. The task of ensuring food security will be quite formidable in an era of increasing climate risks and diminishing farm productivity.

China has already built strong defences against the adverse impact of climate change. In 2010, China produced over 500 million tonnes of foodgrains in a cultivated area similar to that of India. Chinese farmland is however mostly irrigated unlike in India where 60 per cent of the area still remains rain-fed. Food and drinking water are the first among our hierarchical needs. Hence while assessing the common and differentiated impact of a 2°C rise in temperature, priority should go to agriculture and rural livelihoods. What are the steps we should take in the field of both mitigation and adaptation ?

The largest opportunity in the area of mitigation lies in increasing soil carbon sequestration and building up soil carbon banks. Increase in the soil carbon pool in the root zone by 1 tonne C/ha/yr will help to increase food production substantially, since one of the major deficiencies in soil health is low soil organic matter content. There should be a movement for planting a billion "fertiliser trees" which can simultaneously sequester carbon and enhance soil nutrient status. We can also contribute to the reduction in methane emission in the atmosphere from animal husbandry by popularising biogas plants. A biogas plant and a farm pond in every farm will make a substantial contribution to both reducing GHG emission and ensuring energy and water security. Similarly neem-coated urea will help to reduce ammonia volatilisation and thereby the release of nitrous oxide into the atmosphere.

2010 was the International Year of Biodiversity. We can classify our crops into those which are climate resilient and those which are climate sensitive. For example, wheat is a climate-sensitive crop, while rice shows a wide range of adaptation in terms of growing conditions. We will have problems with reference to crops like potato since a higher temperature will render raising disease-free seed potatoes in the plains of northwest India difficult. We will have to shift to cultivating potato from true sexual seed. The relative importance of different diseases and pests will get altered. The wheat crop may suffer more from stem rust which normally remains important only in peninsular India. A search for new genes conferring climate resilience is therefore urgent. We have to build Gene Banks for a warming India.

Anticipatory analysis and action hold the key to climate risk management. The major components of an Action Plan for achieving a Climate Resilient National Food Security System will be the following:

- Establish in each of the 127 Agro-climatic Sub-zones, identified by the Indian Council of Agricultural Research based on cropping systems and weather patterns of the country, a Climate Risk Management Research and Extension Centre.
- Organise a content consortium for each centre consisting of experts in different fields to provide guidance on alternative cropping patterns, contingency plans and compensatory production programmes, when the area witnesses natural calamities like drought, flood, higher temperatures and in the case of coastal areas, a rise in sea level.
- Establish with the help of the Indian Space Research Organisation (ISRO) a Village Resource Centre (VRC) with satellite connection at each of the 127 locations.
- Link the 127 Agro-climate Centres with the National Monsoon Mission, in order to ensure better climate, crop and market intelligence.

- Establish with the help of the Ministry of Earth Sciences and the India Meteorological Department an Agro-Meteorological Station at each Research and Extension Centre to initiate a "Weather Information for All" programme.
- Organise Seed and Grain Banks based on computer simulation models of different weather probabilities and their impact on the normal crops and crop seasons of the area.
- Develop Drought and Flood Codes indicating the anticipatory steps necessary to adapt to the impact of global warming.
- Strengthen the coastal defences against rise in sea level as well as the more frequent occurrence of storms and tsunamis through the establishment of bio-shields of mangroves and non-mangrove species. Also, develop seawater farming and below sea level farming techniques. Establish major Research Centres for Seawater Farming and Below Sea-Level Farming. Kuttanad in Kerala will be a suitable place for the Below Sea-Level Farming Research and Extension Centre. A major centre should also be established in the Sunderbans area of West Bengal.
- Train one woman and one male member of every panchayat to become Climate Risk Managers. They should become well versed in the art and science of climate risk management and should help to blend traditional wisdom with modern science. The Climate Risk Managers should be supported with an internet-connected Village Knowledge Centre.

A Climate Literacy Movement as well as anticipatory action to safeguard the lives and livelihoods of all living in coastal areas and islands will have to be initiated. Integrated coastal zone management procedures involving concurrent attention to both the landward and seaward site of the ocean and to coastal forestry and agro-forestry as well as capture and culture fisheries are urgently needed. A genetic garden for halophytes is being established at Vedaranyam in Tamil Nadu. A very visionary decision has been taken to establish five genetic heritage gardens based on Tamil *Sangam* literature, which 2000 years ago described five major agro-ecological zones in the State, namely, *Kurinji* (hill areas), *Mullai* (forests), *Marudham* (pastoral lands), *Neithal* (coastal areas), *and Palai* (dry lands).

Hereafter, climate care and resilience must be mainstreamed in all development programmes. We can then ensure food, water, fodder and food security for a human population of 1.2 billion and 1 billion farm animals.

❑❑❑

Chapter 26

Maximising the Benefit of A Good Monsoon

A climate-resilient agriculture, which we need urgently, will have to be based on a two-pronged strategy: maximising farm productivity and production during a normal monsoon period, and minimising the adverse impact of unfavourable weather as witnessed during 2009. Unfortunately, we are yet to develop an anticipatory research and extension programme to minimise damage during unfavourable monsoon periods. For example, the deficiency in rainfall during the southwest monsoon of 2009 was 23 per cent. The growth in agriculture and allied sector GDP was minus 0.2. The highest growth rate in agriculture GDP of 5.2 per cent was observed during 2005-06 when the growth in total GDP was 9.5 per cent. Had we had a scientific monsoon management strategy, we could have minimised the loss in 2009-10. Similarly, if we have a strategy for maximising the benefits of a good monsoon, we can hope to achieve at least 5 per cent growth rate during 2010-11 in agriculture and allied sectors. In parts of China like the Yunnan province, which has experienced 60 per cent less rainfall than normal, there is a move to grow different crops together in the same field, thereby distributing the risk arising from monoculture. It is time to develop a pro-active monsoon management strategy.

First, we must improve soil health and help farmers to benefit from the nutrient-based subsidy regime which has been introduced with effect from 1 April 2010. If used properly, this revised approach to fertiliser subsidy should promote balanced fertilisation with concurrent attention to both macro- and micro-nutrients as well as soil organic matter. To benefit from this revised approach, farmers should have access to Soil Health Cards containing credible information on the chemistry, physics and micro-biology of their soils. Some States like Gujarat have started the practice of empowering farm families with Soil Health Cards. Factor productivity is low now because of lack of attention to micro-nutrients and soil organic carbon content.

Second, we must maximise the benefits of all available water sources — rain, ground, river, treated effluents and seawater. The lessons learnt from the more than 5,000 farmer participatory projects organised by the Union Ministry of Water Resources to maximise yield and income per drop of water should be extended to all farms. Every farm in rain-fed and dry farming areas, which constitute 60 per cent of our total cultivated area, should have a farm pond, a biogas plant and a few fertiliser trees. This will help to build Soil Carbon Banks and also farm-level Water Banks, which will help to undertake crop life-saving irrigation when needed. Energy management is another important requirement for irrigation water security. Electricity or diesel is essential for both groundwater use and for lift irrigation.

Third, we should launch a programme for spreading the best available technologies, including the most appropriate seeds in all the agro-climatic zones of our country. The faculty and post-graduate students of agricultural and animal sciences universities, the staff of the various departments of government related to agriculture and irrigation, and representatives of lead banks and NABARD should go from village to village in each one of these zones for checking whether seeds and other essential inputs are available.

Fourth, both credit and insurance agencies should do their best in taking credit to the last mile and last farmer and get them the benefit of the 5 per cent interest rate for farm loans announced by the Finance Minister. Similarly, insurance companies should deliver the benefits of insurance to every farm family. Under the Mahila Kisan Shasakthikaran Pariyojana announced by the Finance Minister, all women farmers should be enabled to have access to credit, technology, inputs and market.

Finally, the economic viability of farming will depend upon access to assured and remunerative markets. The Minimum Support Price announced for nearly 25 crops must be enforced. A national grid of grain storages starting with the Pusa Bin at the farm level, storage godowns at the village level and modern silos at the regional level should be established without further delay. It is painful to observe the spoilage caused to foodgrains as well as to perishable commodities due to poor storage conditions, when we consider the toil of farm women and men in sun and rain to produce them.

Many of our problems in the field of food and nutrition security are not related to the lack of schemes, but to the over-abundance of disjointed programmes operated by different Ministries. For example, there is little coordination among large national programmes like the Rashtriya Krishi Vikas Yojana, Food Security Mission, Horticulture Mission, Mahatma Gandhi National Rural Employment Guarantee Programme and many other projects. If only there is convergence and synergy among these programmes, our progress in improving the productivity, profitability and sustainability of small farm agriculture will be fast.

❑❑❑

Chapter 27

Building Sustainable Water Security Systems

How can we develop a sustainable water security system in an era of climate change and global warming ? In my view, every country should develop a sustainable water security system with the following four major components.

Supply augmentation

The major sources of irrigation water are from rainfall, rivers, tanks, reservoirs and other surface water resources, groundwater, industrial and domestic effluents, and seawater. Seawater comprises almost 97 per cent of the global water resource and is an important social asset. With the melting of Arctic and Antarctic ice, sea level will go up with disastrous consequences. The tsunami of December 2004 gave a glimpse of what could happen in the future. To augment supplies, we must harvest rainwater and store it carefully, both above and below ground. Also, we should ensure that all waste water which emerges from industry and domestic use is purified and recycled. Rainwater harvesting should become a way of life, so that it becomes everybody's business.

Seawater is an invaluable resource for agriculture and aquaculture. Seawater farming can become a normal activity of coastal communities. Seawater farming involves an integrated approach to agroforestry, horticulture, aquaculture and marine fisheries. MSSRF is establishing near Chidambaram in Tamil Nadu, a research and capacity building centre for seawater farming and another in Kuttanad in Kerala for below sea level farming. There are a large number of halophytes like *Salicornia, Atriplex*, etc, which can be grown along the coast. A combination of casuarina, cashewnut and coconut can provide good income to coastal families. At the same time, coastal ecosystems should be strengthened by rehabilitation of degraded mangrove ecosystems.

Thus, by harvesting rainwater, managing the aquifer sustainably, recycling all waste water and industrial effluents, fostering the conjunctive use of ground and surface water, and above all, promoting seawater farming, a sustainable National Water Security System can be developed.

Demand management

Unfortunately, the water resource is generally measured only in quantitative terms. There is not equal interest in getting the best out of the available water by emphasising economy and efficiency of water use. Based on the concepts developed at the International Water Management Institute (IWMI), we launched in 2007, a Farmer Participatory Action Research Programme (FPARP) in India which, with the active involvement of farm families, is spreading technologies which can help to increase yield and income per drop of water. The technological interventions introduced under this programme for maximising the benefits of the available water include:

- System of rice intensification
- Micro-irrigation with fertigation
- Soil health enhancement to promote soil-water synergy
- Promotion of crop-livestock integrated farming systems
- Crop diversification and multiple use of water
- Weather-based crop insurance programmes
- Credit, insurance and market reforms

There are great opportunities for minimising demand through increased water use efficiency in the agricultural and industrial sectors. We should launch a Water Literacy Movement using modern information and communication technologies to make economy and efficiency the bottom line of water use policies.

New technologies

Among new technologies, solar desalination of seawater and the use of genetic modification techniques for developing crop varieties resistant to salinity, drought and flood are important. Solar desalination is still expensive, but with the current emphasis on a better solar future, the cost is likely to come down. Other techniques like reverse osmosis are equally expensive. Nevertheless we have to develop methods of utilising seawater both directly for human needs, as well as indirectly through seawater farming.

As regards biotechnology, there are controversies about biosafety and bioethics issues. These can be resolved by developing transparent and professionally credible methods of risk-benefit assessment. Every country should develop a statutory Biotechnology Regulatory Authority which inspires public, political, professional and media confidence. The Government of India is in the process of enacting legislation for this purpose.

Without water security, the goals of "food, health and work for all" cannot be achieved. Therefore, every nation should develop an implementable and affordable water security system with concurrent attention to supply augmentation, demand management and harnessing of new technologies. We should take advantage of new technologies for harnessing the hitherto untapped opportunities for raising crops under adverse conditions. Above all, we should take anticipatory action to insulate the human and animal population from the adverse impact of global warming.

Anticipatory action to mitigate the impact of global warming

Climate change adaptation and mitigation strategies should include the development of Drought, Flood and Good Weather Codes. The Drought and Flood Codes will provide detailed guidelines to local communities on the steps to be taken for safeguarding lives and livelihoods during such emergencies. A reliable source of water for domestic use and for livestock will have to be developed. The different codes involve the preparation of contingency plans and alternative cropping strategies to match different weather models. The plans have to be backed up with seeds of alternative crops. Just as grain reserves are essential for food security, seed reserves are essential for crop security.

Groundwater sanctuaries, which are concealed aquifers, can be developed for use when absolutely essential. It is unfortunate that traditional irrigation systems in the form of tanks are in a state of decay. Cropping patterns are also changing, because of changing weather patterns. There is therefore need to revisit the management models for surface irrigation systems.

Good Weather Codes will help to maximise agricultural production during normal rainfall seasons. Thus there is need for a twin strategy which will help to minimise the adverse impact of unfavourable weather and maximise the benefits of good seasons.

❑❑❑

Chapter 28

Drought Management for Rural Livelihood Security

There are reports in financial newspapers that the ongoing (2010) drought affecting nearly 200 districts in the country may not have much effect on GDP, since the farmers in the drought-affected areas contribute hardly 3 per cent to GDP. It is sad that such a measure of the impact of drought on the lives and livelihoods of millions of rural families is even considered. It is this mindset, typical of the growing insensitivity to human suffering in our country, that is responsible for India being the home for the largest number of undernourished and malnourished children, women and men in the world. No wonder we are finding it difficult to achieve the first among the UN Millennium Development Goals, namely, reducing hunger and poverty by half by 2015. Unless we realise that agriculture in India is not just a food-producing machine, but is the backbone of the livelihood of over 60 per cent of our population, rural deprivation and suffering will not only continue to persist, but will get worse, leading to severe social unrest.

Fortunately, there are some encouraging developments which offer hope that drought management will be based on human values. First, our President has urged the need for refraining from making profit out of poor peoples' entitlements. This is a timely warning since thousands of crores will be spent during the coming weeks in drought relief. Unfortunately, disaster relief funds become an easy target for those to whom corruption is a way of life.

Second, the Prime Minister has rightly emphasised the need to help farmers in their hour of distress, so that they can help the country to produce as much food as possible under the prevailing meteorological conditions. He has announced that the repayment of loans taken from banks will be rescheduled. In this connection, it will be useful to find a long-term solution to the problems faced by farmers in rain-fed areas by adopting the recommendation of the National Commission on Farmers that the repayment period for loans in drought-prone areas should be 4 to 5 years. This is particularly important, since we do not have an effective crop insurance policy for farmers in drought-prone areas.

Third, a Crisis Management Committee has been constituted under the leadership of Pranab Mukherjee, who fortunately belongs to the rare group of leaders who are firmly rooted in the "We shall Overcome" philosophy. I hope the Crisis Management Committee will not only look into the immediate problems and short-term solutions, but will also develop medium- and long-term plans which can enable us to face the challenges of drought, flood, high temperature and sea level rise which in future will be the recurrent consequences of global warming and climate change. Since serious action involving a large financial outlay is currently under discussion, I would like to lay out a road map on the action needed both immediately and during the remaining period of the Eleventh Five Year Plan.

With the help of State governments, ICAR and agricultural universities, the situation in each State may be classified into the following two categories:

Most seriously-affected areas (MSA) are those where monsoon irregularity has multiple adverse effect on crops, farm animals and human food and livelihood security. Also, hydropower generation is affected, leading to energy shortage. Power shortage in turn makes it difficult to give a crop life-saving irrigation, wherever opportunities for this exist.

Apart from the relief operations normally undertaken, the urgent needs of MSA areas include saving farm animals from distress sale through Farm Animal Camps near a water source or near a groundwater sanctuary (i.e,, concealed aquifer which can be exploited during the emergency) and where animals can be fed with agricultural residues enriched with urea and molasses. Distress sale of farm animals is a clear index of extreme despair.

A "Beyond the Drought Programme" should be organised involving short duration crops like *saathi* maize (60-days maize), sweet potato, pulses, oilseeds, fodder crops and other less water requiring but high value crops, as per scientifically prepared contingency plans.

Another urgent need is the launching of "A Pond in Every Farm" movement, by permitting NREGA workers to build *jal kunds* in the farms of small and marginal farmers. The revised NREGA guidelines permit this. At least 5 cents in every acre should be reserved for the construction of ponds to store rainwater. Where there is adequate groundwater in MSA areas, subsidised electricity and diesel should be made available on a priority basis. Energy is the key limiting factor in taking advantage of groundwater.

Most favourable areas (MFA) can be identified where there is enough moisture for a good crop. A compensatory production programme can be launched in such MFA farms by taking steps to increase the productivity of the crops already sown. This can be achieved by undertaking top-dressing with urea or other needed fertilisers, including micro-nutrients, with government support. Wherever there are opportunities for launching such compensatory production programmes because of adequate rainfall, the faculty and scholars of the agricultural university in the area can be requested to move from classrooms to farmers' fields to help in ensuring the proper administration of the nutrient top-dressing programme. This will help to significantly increase crop productivity.

Where two or more crops are taken normally, it is time to begin preparation for a good *rabi* crop by assembling the seeds, soil nutrients and other agronomic inputs needed for timely sowing and good plant population. Late sowing of *kharif* crops should not be encouraged, since every week's delay in the sowing of wheat reduces the yield by over 4 quintals per hectare.

Detailed Drought, Flood and Good Weather Codes should be prepared for every agro-climatic zone in the country. These codes should indicate the pro-active measures to be taken such as building Seed Banks of alternative crops, needed for minimising the adverse impact of rainfall abnormalities. The Good Weather Code should provide guidelines for maximising the benefits of good soil moisture. Another step urgently needed is something I have emphasised again and again: the identification and training of one woman and male member of every panchayat as Climate Risk Managers. It is best that they are identified by the Gram Sabha. The Climate Risk Managers can be trained in the science and art of managing uncertain rainfall patterns leading to drought or flood. They could also operate a Weather Information for All programme based on village-level agro-meteorological stations. A mini agro-met station can be built in every block with basic instruments to measure temperature, rainfall, wind speed and relative humidity. The Climate Risk Managers can be trained in data collection and interpretation, so that the right decisions are taken at the right time and place. Such a technological upgrading of agricultural infrastructure would also help to attract youth in farming.

In 1966, the country faced a severe drought. A serious famine was avoided, particularly in Bihar, though concessional wheat imports of the order of 10 million tonnes under the US PL-480 programme. This served as a wake-up call and several steps were taken under the far-sighted political leadership of C. Subramaniam, Lal Bahadur Shastri and Indira Gandhi, which led to a wheat revolution in 1968. The major ingredients of this revolution were: technology, services to take the technology to the fields of small and marginal farmers, public policies (particularly relating to input and output pricing), assured and remunerative marketing, and, above all, farmers' enthusiasm as a result of national demonstrations in small farmers' fields. Today, the last component of the Green Revolution symphony is sadly lacking. Over 40 per cent of the farmers interviewed by NSSO said that they want to quit farming, if there is another option. No further time should be lost in implementing the commitments made under the National Policy for Farmers presented in Parliament in November 2007, if the desire of the Prime Minister that there should be another Green Revolution is to materialise.

□□□

Chapter 29

Privatisation of Food and Water Security Systems: An Unequal Social Bargain

Managing the National Food Security System

Rajiv Gandhi understood the wisdom of Gandhiji's doctrine that Gram Swaraj is the pathway to Purna Swaraj and launched the Panchayati Raj movement. Schedule 11 of Constitution Amendment 73 entrusts panchayats with the responsibility of managing natural resources and fostering sustainable agriculture. Representative democracy through elected members, one third of whom are women, and participatory democracy through Gram Sabhas, are powerful tools for ensuring a pro-nature, pro-poor, pro-women, and pro-livelihood orientation to all rural and agricultural development programmes.

The National Commission on Farmers has emphasised the need for community-managed food and water security systems promoted with the help and oversight of Gram Sabhas. The Gram Sabha can serve as a Pani Panchayat to ensure that rainwater is not only harvested, but is used in a sustainable and equitable manner. The trend now is to bypass these grass-root democratic structures, which represent the vision of Rajiv Gandhi for a peaceful and prosperous India, and to resort to depending on private trade to manage our food and water security systems. Unfortunately, this will result in an unequal social bargain since those who control the marketplace are both rich and politically powerful. The Indian enigma of the coexistence of great technological and intellectual capability on the one hand, and extreme poverty, deprivation and malnutrition on the other will continue to persist if we do not revitalise and empower grass-root democratic institutions.

In a country with a high prevalence of poverty and malnutrition, the Government of India should always retain a commanding position in the management of the food security system. This will call for a grain purchase policy which takes into account the changes in the cost of production, (such as a rise in diesel price) subsequent to the announcement of a Minimum Support Price. Traders will give a price above MSP when they expect that prices will shoot up with in a few months. As Professor Amartya Sen has often stressed, we should not forget the lessons of the Bengal Famine of 1942-43, where millions died out of starvation not because there was no food in the market, but because the surplus food stocks were in the hands of private merchants. Building a sustainable food security system will require attention to both the availability of sufficient stocks and who controls them. The global wheat stocks are down and the political leadership of the country should decide how to ensure the food security of 1.1 billion children, women and men in an era where much of the foodgrain stocks will be controlled by national and international grain traders and cartels.

Launching a second Green Revolution

The year 1968 marked the beginning of the first Green Revolution when Indira Gandhi released a special stamp titled "Wheat Revolution". Green Revolution implies enhancing food production through raising productivity per units of land, water, time and labour. The productivity pathway is the only one available to population-rich but land-hungry countries like ours for achieving a balance between human numbers and food production. Even as far back as 1968, I had called for the mainstreaming of environmental concerns in agricultural research and development to avoid the adverse ecological consequences of exploitative agriculture. Later, I coined the term "ever-green revolution" to indicate the pathway to improving productivity in perpetuity without associated ecological or social harm. This term has now come into widespread use internationally.

Prime Minister Manmohan Singh has been calling for a second Green Revolution. It will be appropriate to restrict the use of this term to enhancing the productivity, profitability and sustainability of dry-land farming, i.e., raising crops solely based on rainwater. If the first Green Revolution benefited farmers in irrigated areas, the second should help farm families in rain-fed, semi-arid areas. In both cases, the pathway used for yield enhancement should be the ever-green revolution approach.

If we are to achieve a second Green Revolution covering rain-fed areas, the first important requisite is opportunity for assured and remunerative marketing for dry-land farm products like pulses, oilseeds, millets, vegetables, fruits, milk and meat. This will help to enhance nutrition security on the one hand, and the productivity and economic sustainability of improved dry-land agriculture, on the other. There is a large untapped reservoir of dry-land farming technologies and we can see a drastic rise in the productivity and production of crops in these areas if farm families are supported by credit, insurance, a fair price and assured market for their produce, as happened in the 1980s when Rajiv Gandhi launched a Technology Mission in Oilseeds.

The country can produce as much pulses and oilseeds as we need through a synergy between technology and public policy, since there is a stockpile of improved varieties of dry-land crops. The new hybrid *arhar* strains (pigeon pea) can trigger a pulses revolution. The largest section of consumers in India is the farming population. By helping farmer-consumers to have greater marketable surplus because of higher productivity, we can substantially eliminate poverty-induced hunger and malnutrition in the country.

I have stressed again and again that our food budget should be managed with home-grown food, since agriculture is the backbone of our rural livelihood security system. Importing foodgrains, if continued in the long run, may help some traders and multinational companies to become rich, but will render millions of farm women and men in rain-fed areas paupers.

Another step that should be taken in dry farming and tribal areas is the establishment of community-managed food and water security systems. This will involve the establishment of Grain and Water Banks by local self-help groups. The Grain Bank could be built up with local staples and could help to avoid distress sale as well as panic purchase. The Water Bank can be established by community water harvesting. Conservation, cultivation, consumption and commerce can become an integrated food management system under the control of local communities. By promoting such decentralised community management systems, with the Gram Sabhas providing policy oversight, we can concurrently address endemic hunger caused by poverty, hidden hunger arising from the deficiency of iron, iodine, zinc and vitamin A in the diet, and transient hunger caused by natural calamities like drought, floods, cyclones. It will also be prudent to develop such a system in the context of potential adverse changes in temperature, precipitation and sea level arising from climate change and global warming.

Increasing privatisation of our food and water security systems has important implications for the food, income and work security of small and marginal farmers and agricultural labour. The WTO (World Trade Organization) agreement entered into at Marrakesh in 1994 resulted in an unequal trade bargain. The growing privatisation of food and water security systems is already leading to an unequal social bargain. The poor will not be able to withstand the tragedy of distress sales and inundation of low-cost foods and fruits from rich countries whose agriculture is driven by heavy inputs of subsidy, capital and technology. We will never be able to achieve the UN Millennium Development Goal in the area of hunger and poverty elimination, if we do not insulate the farmer-consumers from unfair trade and social bargains.

A Universal Public Distribution System, which alone can save the economically underprivileged sections of the society from chronic undernutrition, will annually need approximately about 40 million tonnes of foodgrains. If we assume that about 160 million families will use PDS, and that each family gets an allocation of 20 kg per month, we will annually need about 38 million tonnes to support a Universal PDS system. By enlarging the minimum support price to a wider range of foodgrains and purchasing them for use in PDS, we can launch both a second Green Revolution and a universal public distribution system. Also, through Community Grain and Water Banks, we can help to start a "Store Grains and Water Everywhere" movement. The prices of essential commodities will then remain stable and affordable to resource- poor consumers. Market manipulation of prices of essential commodities can also be checked. While foodgrain imports will provide a breathing spell in controlling price rise and inflation, a second Green Revolution in dry farming areas, stimulated by assured and remunerative marketing opportunities, will help to promote food and livelihood security to millions of small and marginal farmers and landless labour simultaneously.

To conclude, import/ export of pulses, oilseeds and wheat may be necessary in years of shortfall or surplus. What is important is to recognise that imports of pulses and oilseeds serve as indicators of our failure to launch a green revolution in dry-farming areas, in spite of having the technologies and resources to do so. This is not just a matter for national pride or shame, but a human tragedy of vast dimensions where millions of children, women and men are condemned to a life of malnutrition and poverty. Imports of crops of importance to the income security of farm families in rain-fed areas imply generating more unemployment and misery in such areas. The mindset where the term "consumers" applies only to the politically powerful urban population and ignores the 70 per cent of the population living in rural India, who are both farmers and consumers, will have to be destroyed if our country is to achieve Purna Swaraj.

❑❑❑

Chapter 30

Shaping the Future of Agriculture in Northeastern India

India's Northeast region — one of natural beauty and cultural richness — is both a mega-biodiversity area as well as a "hot-spot" for genetic erosion. The rural population is around 82 per cent and depends largely on agriculture and allied sectors for work and income security. The forests harbour 8000 out of the 15,000 species of flowering plants occurring in the country. Of about 1300 species of orchids reported from India, the Northeast has the highest concentration with about 700 species. The species richness is highest in Arunachal Pradesh where over 5000 flowering plants occur and the lowest in Tripura with 1600 species. Bamboo is the lifeline of the Northeast and 63 out of 136 species found in India occur in this region. Sadly, however, 25 species of bamboo fall under the rare and endangered category. The region is also home to the *eri* and *muga* silkworms. The yak and the *mithun* are unique animals which are threatened by the spread of non-edible invasive species like *Lantana, Eupatorium* and *Mekenia*. The cultural diversity of this region is well known, with 225 tribes out of 450 in the country living here. In spite of the richness of culture and bio-resources, there is extensive poverty and unemployment. The allocation of funds for the Northeast is quite high and all scientific departments are required to spend at least 10 per cent of their budget in this region. The large outlay is however not getting converted into socially meaningful outcome.

During 1972-75, I set up an ICAR Research Complex for this region with headquarters at Barapani in Meghalaya and with Centres in Arunachal Pradesh, Manipur, Mizoram, Nagaland, Sikkim and Tripura. Agricultural Universities exist at Jorhat and Imphal. The Department of Biotechnology has also set up an Institute of Bio-resources and Sustainable Development at Imphal. ICAR National Research Centres exist for *mithun* in Nagaland, yak in Arunachal Pradesh, pig in Guwahati and orchid in Gangtok. There are many other research institutions including the Northeastern Space Application Centre of ISRO and a Regional Centre of the Indira Gandhi National Open University.

Ecologically, the region has serious problems like the damage caused by deforestation and exploitative mining. The Uranium Corporation of India is trying to convince the people of the West Khasi Hills that uranium mining can be done in an ecologically and socially desirable manner. In spite of an average annual rainfall of over 2000 mm in the region, water shortage is serious during the period December to May. The Cherrapunjee region (now known as Sohra), located in the East Khasi Hills of Meghalaya, experiences a serious water scarcity between December and April, although the annual rainfall exceeds 12,000 mm. Sohra is located on top of a limestone plateau. Limestone sucks up the water. The surrounding hills are denuded and more than 50 per cent of the forests have been lost. The only solution is to hold rainwater where it falls. Every household must have a tank to hold rainwater that can be collected from the roof. In Mizoram, every house collects rainwater in tanks made of tin or concrete situated either on the ground or underground. This water lasts for the post-monsoon months. The entire region can emulate this example for ensuring year-round water security. Fortunately, the Meghalaya Government has initiated steps for promoting a jal kund or farm pond movement.

The region produces about 5.8 million tonnes of foodgrains as against the requirement of about 7.5 million tonnes. The gap between potential and actual yields is high in most farming systems, ranging from 155 per cent in mustard and going up to 650 per cent in wheat. Thanks to the Horticulture Mission, area has increased under most fruits and vegetables. However, the productivity is still low. Hence "bridging the yield gap movement" assumes urgency, since most holdings are small and there is need for greater marketable surplus and cash income. Where flora is concerned, Mizoram produces some of the finest anthuriums in the country and Sikkim is the home of beautiful orchids.

Based on considerations of ecology, economics and employment generation, MSSRF has developed the following approaches to enhance opportunities for sustainable livelihoods in biodiversity-rich areas such as the Northeast:

- *Biovillages.* The Biovillage paradigm involves concurrent attention to a) the conservation and enhancement of the ecological foundations for sustainable agriculture; b) improvement of the productivity and profitability of small holdings, and c) generation of multiple livelihood opportunities through crop-livestock-fish integration, biomass utilisation and agro-processing.
- *BioParks.* Such Parks are designed to add value to plant and animal biomass through agro-processing and preparation of a wide range of market-linked products. A Rice Bio-Park can be established in Manipur.
- *BioValleys.* The aim of the Biovalley is to promote along a watershed small- scale enterprises supported by micro credit and to link biodiversity, biotechnology and business in a mutually-reinforcing manner. The biotechnology enterprises relate to the production and marketing of the biological software essential for sustainable agriculture such as biofertilisers, biopesticides, and vermiculture.

Such institutional approaches will help to foster an economic stake in conservation. The Northeast region is a mine of valuable genes in rice which confer resistance to several biotic and abiotic stresses. The genetic variability in rice includes a strain which is reported to be one of the tallest among the global rice germplasms. It would be useful to undertake jointly with farm families a hybridisation programme involving crosses between such unique rices and modern varieties possessing a high yield potential and desired duration and grain quality. In West Africa, for example, rice yield was stagnant for a very long time, until scientists crossed a West African rice species *Oryza glaberrima* with the Asian rice *Oryza sativa.* The hybrids were then given to farm families who selected appropriate strains suited to local conditions and tastes. These rices are known as New Rices for Africa (NERICA). There is similar scope for new rices for the Northeast, through farmer participatory breeding. Participatory research and participatory knowledge management hold the key to successful and sustainable agriculture in this region.

A high priority in this region is human resource development. Unfortunately, more than 50 per cent of the scientific posts in the existing institutions in the field of agricultural research and education are vacant. Also, many of the positions are held by persons from outside the region, several of whom do not have a long-term stake in linking science with society in the region. There is now a trend to deviate from what is known as the "Bhabha model of nurturing science". Homi Bhabha built institutions around outstanding individuals and established the Trombay School to breed a new class of brilliant scientists and science leaders. Regrettably there is now a reversal of this paradigm with the highest priority going to the brick and mortar aspect of institution building. I have therefore been suggesting that in the case of agricultural science about a thousand women and men graduates should be selected from the region, provided with fellowships to do the M.Sc / Ph.d. degrees in appropriate agricultural and animal sciences universities, and then inducted into ICAR's Agricultural Research Service (ARS). I got ARS established in ICAR in 1974 in order to promote a scientist-centered system in personnel policies in the place of the then prevailing post-centered system. A special Northeast cadre of ARS comprising of over 1000 women and men scientists belonging to the region will help to achieve what the Royal Commission on Agriculture emphasised as early as 1925:

> However efficient the organisation which is built up for demonstration and propaganda be, unless that organisation is based on the solid foundation provided by research, it will be merely a house built on sand.

The sad irony of the coexistence of poverty of people and prosperity of nature will continue so long as the political approach is on quantitative expansion of educational institutions without corresponding emphasis on the breeding of outstanding teachers and researchers from the region. We must shift our emphasis from bricks to brains.

Agricultural Transformation Centres must be initiated in all the blocks in this region and every farm family issued a Farm Health Passbook. Mahila Kisans (Women Famers) and Yuva Kisans (Young Farmers) will determine the future of the agrarian and rural economy of this region. A Mahila Kisan Sashaktikaran Pariyojana has been introduced and I hope a Yuva Kisan

Sashaktikaran Yojana wil follow soon. Challenges are many, but opportunities are equally many. Some crops like rubber may do better in this region than in Kerala when mean temperature goes up by 1-2° C due to climate change. I see a glorious future for the agriculture of this region in an era of climate change, with the prospects of farming and farmers being shaped by harnessing the best in frontier science and blending it with traditional wisdom and ecological prudence.

❑❑❑

Chapter 31

Towards Vidarbha's Agricultural Renaissance

Livestock and livelihoods are intimately linked in our country. Mixed farming involving crop-livestock as well as sometimes fish integration has been both a way of life and a means to sustainable livelihoods. Also, the ownership of the livestock is more egalitarian than that of land. That is why in our ancient culture, livestock has been given a pride of place in the area of spiritual veneration and natural resources conservation. Maharashtra is rich in its livestock resources and has also a long coastline with a wide range of living aquatic resources.

The crop-livestock integrated farming system provides an opportunity for promoting organic farming and restoration of soil health, and thereby of an ever-green revolution leading to an enhancement of productivity in perpetuity without associated ecological harm. It also strengthens household nutrition security. The country as a whole has over 500 million farm animals. However, grazing land has been reduced to less than 5 per cent of the arable land area. Village common property resources, which used to provide fodder and shelter to farm animals, have been overexploited. Consequently, our farm animal productivity is very low.

I wish to deal briefly with a few issues which need to be considered while developing a strategy for a prosperous animal husbandry and fisheries programme. First and foremost, we should deal with the saving of our genetic resources in farm animals and fish. Livestock husbandry is an age-old occupation in Indian agriculture, since farm animals play an important role in nutritional security and sustainable agriculture. Besides providing milk, eggs and meat, they also provide a diverse range of services in areas such as manure, irrigation, transportation, and fibre and leather goods. The value of output from the livestock sector in India, without considering their indirect contributions, is of the order of Rs. 200,000 crore, which is about 30 per cent of the value of the agricultural sector output. Livestock farming is the mainstay of the economy of many of the world's harshest environments — deserts, steppes, mountains, Arctic and Antarctic regions.

The Indian domestic animal diversity is rich with more than 30 breeds of cattle, 10 breeds of buffalo, 42 breeds of sheep, 20 breeds of goat, 8 breeds of camel, 6 breeds of horse and 18 breeds of poultry. Numerous breeds of cattle have been domesticated based on local needs and their adaptability under different eco-geographical and agro-climatic regions. Some are known for their draught power, some for milk and some for dual purposes. Similarly, wide genetic variability exists among sheep and goats.

India is among the very few countries having such a large number of breeds of farm animals with a wide genetic diversity. Livestock-integrated agriculture is unique because of its diversity and location-specific requirements. India has contributed richly to the international livestock gene pool and improvement of animal production in the world. For example, *Brahman* cattle are found in 45 countries, while the *Sahiwal* breed is found in 29 countries. Many cattle breeds of Indian origin have made major contributions to the development of composite breeds elsewhere in the world, as for example, Brazil.

Due to changing land use pattern in agriculture, many of the breeds are either showing a declining trend or are on the verge of extinction. Many are also facing genetic deterioration due to the shrinking of grazing lands, ineffective and unscientific breeding programmes, increased mechanisation and less emphasis on livestock-based livelihood systems. As a result, Indian livestock breeds, once famous for their draught capacity, heat tolerance, disease resistance, and adaptability to harsh agro-climatic conditions, are undergoing genetic erosion. This calls for urgent attention to conserving domestic animal diversity.

There is higher biodiversity in cattle, goat and sheep than in pig, fowl, quail, geese, ducks, mithun, yak and pet animals. Many of the native breeds of these livestock possess remarkable ability to resist endemic diseases and to subsist on local feed and fodder resources. Their maintenance cost is relatively low and hence they are more suitable to small farm holdings. Conservation of these traits and breeds are important for future agriculture, which may have to face new challenges arising from global warming.

Among the livestock diversity, the diversity of buffaloes in India is very unique. Buffalo breeds such as *Murrah, Nil Ravi, Surti, Mehsana*, and *Jaffarabadi* are notable for their economic value. The *Badawari* breed of buffalos has over 10 per cent butter fat content in milk. Among sheep, *Bikanery, Hissar Dale, Chokra, Mogra, Nali, Marwari, Sonadi* and *Kathiawari, Nellore, Bellary, Mandya, Sahabady*, etc., are prominent. *Angora, Pashmina, Kashmiri, Gaddi, Jamunapari, Barbari Beetal, Berari, Surti*, etc., represent high diversity among the goats that contribute immensely to the economy of the poor in fragile environments. The *Sahiwal, Tharparkar, Gir, Red Sindhi. Amrithmahal, Hallikar, Khillari, Kangayam, Ongole, Deoni* and *Krishanavalley, Punganur, Malnad Gidda, Vechur, Bargur, Umblacherry, Alambadi* breeds of cattle constitute a rich diversity of livestock resources. Unfortunately, for different reasons, population of these important breeds is on the decline. This therefore calls for urgent intervention to arrest the trend and promote conservation efforts at the grass-roots level, with active community participation.

In an era of climate change, the conservation of these genetic resources of livestock is of paramount importance. The conservation of livestock genetic resources should not be only for preserving and maintaining the existing breeds, but also to develop proper management and genetic improvement of indigenous breeds to enhance the diversity and identify economically important genes and gene clusters. As a huge part of the variability is created and conserved by livestock-rearing farmers, it is relevant and appropriate that any conservation effort should start at farmers' level with an agro-ecological reach. Identification of farming communities conserving different major breeds and groups of livestock and rewarding and recognising their contribution economically and socially would be an important initiative for promoting the conservation of livestock genetic resources.

It is therefore essential to recognise and protect the rights of the communities / farmers with regards to their contributions to conserving, improving and utilising livestock germplasm in the development of improved breeds. It is in this context that the National Bioresource Development Board (NBDB) has taken the initiative in introducing a Reward and Recognition System for recognising and rewarding the past and present contributions of the livestock breed conservers. I hope the establishment of a Breed Saviour Award will help to stimulate interest in the conservation of our rich genetic diversity in farm animals, which can help in creating genetic literacy with reference to animal wealth in our villages.

Another aspect which needs attention is the growing conflict between farmers and wild animals. Crop-damaging *rohi*s (*nilgai*) have been giving sleepless nights to farmers in Vidarbha villages, particularly in those adjoining forest areas. The fields close to the forest area have become breeding grounds for *rohi*s as plenty of water and food are available for these animals. There is very little compensation for the damage done by wild animals to crops. Farmers are using traditional methods like putting up scarecrows. Some use light to prevent the wild animals coming to the fields. Some farmers dig trenches around the fields, so that the animals are unable to enter the field. What is now important is a good scientific study of methods of promoting harmony between wild life and human beings. This will be possible only if the habitats of wild life are protected.

Another area of importance is the development of management methods which can give the power and economy of scale to small-scale animal husbandry enterprises. This will be possible by giving centralised services to support decentralised production. We have to combine the economic advantages of mass production technologies with the ecological and social advantages of production by masses. Gandhiji, who made Sevagram the unofficial capital of India, used to mention often that the goat is a poor man's cow. He wanted a pro-poor and pro-woman orientation to livestock enterprises. Nearly 75 million women are now involved in producing over 100 million tonnes of milk annually, thereby making India the leading milk producer of the world. In all perishable commodities, the most important need of farmers is assured and remunerative marketing through processing and value addition. This is where we are indebted to the National Dairy Development Board for its monumental contributions to promoting dairy cooperatives and to linking conservation, rearing, consumption and commerce in a mutually reinforcing manner.

Maharashtra has immense opportunities for both capture and culture fisheries. The Government of India has recently proposed an integrated coastal zone management procedure which takes into account both the land and the sea for conservation and sustainable use. The coastal zone has been defined in the Government notification as the area from the territorial waters limit (12 nautical miles measured from the appropriate baseline) including its sea bed, the adjacent land area along the coast and inland water bodies influenced by tidal action including its bed, up to the landward boundary of the local self government or local authority abutting the sea coast, provided that in case of ecologically and culturally sensitive areas, the entire biological or physical boundary of the area may be included as specified under the provisions of Environment Protection Act, 1986.

A mean sea level rise of 15 to 38 cm is projected along India's coast by the mid twenty-first century and of 46 to 59 cm by 2100. An increase in the frequency and intensity of tropical cyclones is likely. Protective measures include construction of coastal protection infrastructure and cyclone shelters as well as plantation of coastal forests and mangroves. We should develop an integrated coastal zone management plan for Maharashtra involving the development of mangrove and non-mangrove bioshields, including the planting of vetiver. There is also need to safeguard the traditional rights of fisher communities. We already have the Scheduled Tribes and other Traditional Forest Dwellers (Recognition of Forest Rights) Act of 2006, to protect the interests of tribal families depending on forests for their food, medicines and livelihood. I suggest that we should enact a Fisher Communities and Traditional Coastal Dwellers (Recognition of Sea Farming Rights) Act in order to ensure that industrialisation and commercial activities do not destroy the livelihoods of fisher families, whose sole livelihood source is the ocean.

I would also suggest the promotion of aquaculture estates to help small-scale fisher families to have access to good seed and nutritious feed as well as producer-oriented marketing opportunities. There is also need to establish a Fish for All Training Centre which can help to train fisher women and men in all aspects of fisheries ranging from capture / culture to consumption based on the pedagogic methodology of learning by doing.

Modern science and technology have opened up great opportunities for achieving the goals of food, education, health and work for all. Rashtrasant Tukdoji Maharaj has pointed out in his *Gramgeeta* that 'India is a land of glaring and almost stupefying contradictions; yet there is unity in diversity.' The *Gramgeeta* provides guidelines for a happy, healthy and satisfying life. There is a whole chapter in *Gramgeeta* on cattle improvement. According to Shri Tukdoji Maharaj, nature will surely offer health and wealth to all villagers provided we conserve our natural resources and use them in a sustainable and equitable manner.

Let us combine the best of traditional wisdom with modern science and technology including biotechnology, space technology, information and communication technology, nuclear technology and renewable energy technology, in order to develop ecotechnologies rooted in the principles of ecology, economics, social and gender equity, energy conservation and employment generation. Only a combination of the ecological prudence of the past and modern science will help us to achieve rural prosperity and the birth of a new Vidarbha.

❑❑❑

Chapter 32

Science's Role in Stimulating West Bengal's Agricultural Performance

Mahatma Gandhi mentioned to those who wanted to serve rural India that their most useful contribution would be in linking brain with brawn. Only such a combination of intellect and labour can lead to sustainable advances in the productivity and profitability of agricultural holdings. West Bengal's Bidhan Chandra Krishi Vishwavidyalaya (BCKV), almost four decades old, is a fitting example of such work. It has helped the development of two more universities in the State, namely, West Bengal University of Animal & Fishery Sciences and the Uttar Banga Krishi Viswavidyalaya. It is important that these three universities function like a consortium of educational institutions mandated to help in achieving agricultural advance, agrarian prosperity and rural transformation in the State.

Livestock and livelihoods are intimately related in our country. The ownership of livestock is more egalitarian than that of land. Similarly, West Bengal is full of fish ponds. It is important that cultivation of crops like rice is done by adopting integrated pest management procedures, so that the water in which fish is cultivated does not get polluted with pesticides. Thus synergetic linkages among crop husbandry, animal husbandry and fisheries make it imperative that the farmers are given holistic advice based on an entire farming system. Disciplinary depth will have to be combined with an inter-disciplinary problem-solving approach.

Looking back, it might have been more prudent for us to have a single united Farm University in every State, giving concurrent attention to crop husbandry, horticulture, animal husbandry, culture and capture fisheries, agro-forestry and agro-processing. Such an Integrated University could have several autonomous campuses, as for example, the University of California system which includes several autonomous campuses like those at Davis, Berkeley, Los Angeles, etc. While I know this kind of integrated approach is becoming politically more difficult, at least the Vice Chancellors of the different universities could develop working procedures by which the farm and fisher communities are empowered with necessary technical knowledge and skills on a farming system basis. The over-arching goal should be the sustainable livelihood security of farm, fisher and tribal families and the ecological security of rural areas.

The net sown area in West Bengal is about 61 per cent of the total area. In India as a whole by comparison only 46 per cent of the total area is under cultivation (2003-04). Also the share of fallow land, unculturable land and pastures is very low in West Bengal, amounting only to 1.2 per cent of the land under different uses (2005-06). Much of the barren and uncultivable land is concentrated in the districts of Paschim Medinipur, Purulia, Bankura, Darjeeling and Barddhaman. Thus no more land is available in West Bengal for being brought under the plough. Land holdings are getting smaller and smaller each year. The smaller the holding the greater is the need for higher marketable surplus and multiple livelihood opportunities. Redistributive land reform has helped West Bengal to enhance the production and productivity of major crops and thereby reduce poverty. Major beneficiaries of the distribution of the agricultural land have been Dalits, Muslims and Adivasi households. The socially underprivileged communities have gained access to agricultural and homestead land through the process of land reforms. This is a significant step in eliminating poverty and deprivation.

Poverty eradication can be achieved only through asset building. One form of asset is land, livestock, fish ponds, etc. The other form of asset is knowledge, skills and ability to add value to primary products through agro-processing and agri-business. Technical skills help to add value to time and labour. Agricultural universities have to play a major role in attracting and retaining youth in farming. Hence I will like to deal with this issue briefly.

Attracting and retaining youth in farming

Indian agriculture is at demographic, ecological, economic and technological cross-roads. Demographically, nearly 70 per cent of the population is below 35 years in age and 70 per cent of them live in rural areas. Within the next few decades, India will have the world's largest number of young people looking for employment. Ecologically, the basic life support systems of land, water, forests, biodiversity and climate are in distress and the population-supporting capacity of major ecosystems is being exceeded in most parts of the country. India's ecological debt and nature deficit disorder are growing. Economically, the cost-risk-return structure of farming is becoming adverse, with the result that over 40 per cent of today's farmers would like to quit farming, if they had another livelihood option. Technologically, there is a serious mismatch between production and post-harvest technologies. Also, there is no nationally agreed policy towards the technological transformation of agriculture, involving farm mechanisation and frontier technologies like recombinant DNA technology and nano technology.

The recent global economic crisis and violent undulations in fossil fuel prices have underlined the importance of building our food security system with home-grown food. Agriculture is serving as the saviour in the context of the global economic and employment crisis. While industry promotes job-less economic growth, agriculture fosters job-led growth. A study in the *Economic and Political Weekly* has concluded that the jobless growth performance of India's organised manufacturing sector is a matter of serious concern. The economic revival package should therefore lay special stress on the building of a national grid of warehouses, rural godowns and other storage structures both for grains and perishable commodities. A national grid of storage structures will help to maintain price levels and prevent both panic purchase and distress sales. Greater investment in rural infrastructure will help to enhance agrarian prosperity.

While developing strategies for youth involvement in agriculture, it will be necessary to tailor them for the following categories.

- Young women and men operating their own farm (farmers with large and small holdings will need different types of assistance)

- Youth educated in agriculture and allied enterprises, who will be able to organise advisory services, provide equipment and implements on a custom-hire basis and operate agri-clinics and agri-business centres
- Landless agriculture labour whose children can be trained to manage farms in land taken on lease and as well as work in non-farm enterprises
- Educated youth from urban areas who are interested in promoting urban and rurban agriculture, including greenhouse horticulture
- Young farmers who will be able to operate Farm Schools in their farms in order to promote farmer-to-farmer learning

The ultimate aim in all the cases should be the creation of opportunities for meaningful livelihoods for youth in the farm sector, including farm-based non-farm enterprises.

Technology and training

The technology divide is an important cause of the North-South economic divide. Technologies which can help to add economic value to the time and labour of farmers and farm labour can help to alleviate poverty and deprivation. This is why Jawaharlal Nehru made the following statement over sixty years ago:

> It is science alone that can solve the problems of hunger and poverty. The future belongs to science and those who make friends with science.

We have over 60 Agricultural, Animal Sciences, Fisheries, Forestry, Rural and Women's (Home Science) Universities. They can play a leading role in instilling in their alumni the self-confidence and capability essential for taking to agriculture as a profession. This will call for both intensive practical training through institutional mechanisms like Shiksha Dairy (Anand) and Fish for All Training Centre (Poompuhar), as well as a counselling service in the university which can perform a hand-holding role throughout the career of the student. The existing placement bureaus should have a separate window for self-employment. There are many opportunities today to take to a promising self-employment career through agri-clinics, agri-business centres, Food Parks, Biotech Parks, etc. The government also should decide to involve at least one home science graduate with expertise in the field of nutrition in introducing horticultural remedies to the nutritional maladies prevailing in every block. This can be done under the National Horticulture Mission. Group financing and group insurance should be done to stimulate such clusters of expertise being formed to render up-to-date and effective advice to farmers at the production and post-production stages.

The graduates of Farm Universities can either take to farming itself as a profession or organise consultancy services to bring about a technological upgrading of farm activities. Universities should also strengthen their infrastructure and training capability in post-harvest technology and trade, particularly in the case of vegetables, fruits and flowers. As suggested by the National Commission on Farmers (NCF), every scholar should become an entrepreneur. This will call for restructuring the curriculum in such a manner that both scientific discipline and business management principles are taught together — as for example, seed technology and business. The government should also provide social recognition to farm graduates by allowing them to be designated as Farm Practitioners as in the case of Medical Practitioners. A suitable certificate will have to be given for this purpose after careful screening. In order to attract rural students to take to agricultural university education, there should be more Fellowships and "learn while you earn" programmes.

A pro-nature, pro-small farmer, pro-woman and pro-livelihood orientation should be the bottom line of all technology development and delivery programmes.

Small farm management

A large majority of the farms are less than 2 hectares in size. Youth will be attracted to farming as a profession under conditions of small holdings only if the productivity and profitability of such farms can be enhanced through synergy between technology and public policy. Institutional structures like seed villages, biovillages, bioparks and biovalleys and greenhouse horticulture will help scholars to introduce science and technology in day-to-day farm life. At least one woman and one male member of every panchayat should be trained in as Climate Risk Managers in order to help the village to handle situations like drought, flood and sea level rise in an effective manner. A small farm is ideal for ecological agriculture, but a small farmer suffers from many handicaps arising from lack of appropriate and timely services. This is where farm graduates can play a critical role in organising an integrated agri-clinic- cum-agri-business centre in every block. In order to provide multiple sources of income, it will be essential to promote mixed farming involving crop, livestock, fish, agro-forestry and agro-processing. Therefore, there is a strong need for a unified knowledge delivery system in rural areas.

It will be appropriate if the various government subsidies go directly to farmers rather than to fertiliser companies and other agencies as is being done at present. The Government of India has already taken the initiative for transferring fertiliser subsidy directly to farmers on an experimental basis in one district of every State. This can be done by developing a Smart Card method of entitlement like the Kisan Credit Card.

There is also need for proper certification of inputs like seeds, pesticides, vaccines, etc. Young farmers could be trained in the production of the biological software essential for sustainable agriculture, such as biopesticides, biofertilisers and vermi-compost. Decentralised production of hybrid seeds by young farmers will help to reduce the cost of seeds. In the case of aquaculture, there is need for certifying seed and feed, so that there is quality assurance. Many such services can be provided by educated young farm women and men. Protected agriculture in greenhouses can help to popularise drip irrigation and fertigation methods.

Women farmers should be assisted with appropriate implements, including low H.P. tractors. Constraints in the issue of Kisan Credit Cards to women farmers should be removed. Appropriate support services like crèches should be provided to working mothers. The MSSRF Biovillage model of human-centered development is ideal for West Bengal, since it involves concurrent attention to ecological security and to sustainable livelihoods based on higher on-farm productivity and market-linked non-farm employment.

Urban and rurban agriculture (metropolitan agriculture)

With increasing urbanisation, opportunities exist for linking the village and the town in a symbiotic manner from the point of view of producing commodities for which there is a ready market in the town or city. Several parts of our country like Kerala and the Punjab are already rurban in character, with the town and the village forming a continuum. The experience of the Netherlands in promoting Metropolitan Agriculture could be taken advantage of in developing India's urban agricultural programmes. This will provide an opportunity to link rural agricultural entrepreneurs to metropolitan markets. The diversified metropolitan consumer market is an excellent driver for innovations aimed at high quality food, contributing to a better quality of life. Such rural-urban linkages will also help to establish market-linked agro-food networks and thereby make the present Government of India's programmes like Food Parks and Super Food Parks a success. At present, many of these Parks are not functioning well due to lack of backward linkages with technology and raw material, and forward linkages with markets. The work done by the Bharat Chamber of Commerce in the Kulpi block of the Sunderbans in linking the rural producer and the urban consumer needs to be replicated.

Public policy support

Our ability to attract and retain youth in farming will largely depend on the economics of farming. At the moment, most farmers with small holdings find that agriculture alone cannot give them a living wage. Multiple sources of income become important for economic survival and for escaping a deadly debt trap. Credit, insurance and venture capital policies must be restructured for enabling scholars to take to farm enterprises. The following areas need particular attention.

- Establishment of a Farmers' Income Commission to review the net income of farmers and suggest methods of ensuring a minimum income. The aim here is the same as that of the VI Pay Commission for government employees. The Farmers' Income Commission should recommend public policy measures needed to make farming a profitable enterprise.
- Establishment of a National Bank for Farm and Food Security: NABARD has multiple functions and the time has come for the establishment of a Bank which will concentrate entirely on the needs of farmers and of national food security.
- Revitalisation of Land Use Boards in order to equip them to provide pro-active advice to farmers on land use based on meteorological forecasts, irrigation water availability and market demand, both in

home and external markets.

- Establishment of a National Network of Village Resource Centres and Village Knowledge Centres in order to provide farmers with dynamic and location-specific advice relating to both the production and post-harvest phases of farming. Farm graduates may be given priority in the allocation of the Community Service Centres (CSCs) under the programme initiated by the Department of Information Technology. They will then be able to make CSCs more relevant to the needs of the local population and help to impart quality, trade and genetic literacy. Genetic literacy is particularly important with reference to genetically modified crops.
- Engendering all public policies and support services to young women farmers, so that women farmers are able to make an active contribution to our agricultural renaissance.

By 2030, India's population will exceed that of China. Agriculture will continue to remain the backbone of our livelihood security system. Also, farmer-consumers will continue to constitute the majority of consumers. Therefore we will not be able to overcome the prevailing widespread malnutrition, if we do not assist the farmer-consumer to consume more. We should hence enable our farm youth including farm graduates to impart an income and employment orientation to farming, as stressed in the National Policy for Farmers. Agriculture should yield more food, more jobs and more income if we are to achieve the goals of food and work for all.

The youth and the poor are the two genuine majorities of our country. Youth empowerment is the pathway for overcoming the problems caused by poverty and deprivation. In this task, BCKV has to play a catalytic role keeping in mind Dr. B. C. Roy's words:

> We have the ability and if, with faith in our future, we exert ourselves with determination, nothing, I am sure, no obstacles, however formidable or insurmountable they may appear at present, can stop our progress.

❑❑❑

Chapter 33

Singur and Our Socio-Economic Future

Mamata Banerjee has raised the important issue of land use and acquisition for public, political, professional and media debate. With increasing population pressure on land, this issue deserves careful consideration and rational discussion. The Singur dilemma in a wider sense relates to both land use policy and norms of compensation for land acquired by government for non-farm purposes. Land conflicts are also spreading in other parts of our country, as far example where land is needed for establishing Special Economic Zones (SEZs). The Maharashtra Government has wisely adopted the policy of holding a referendum to seek the views of farmers and others who will be dispossessed of their land for meeting the needs of SEZs. Pro-active consultations will help to avoid difficulties later. Water conflicts are also affecting important irrigation projects. Land conflicts are likely to do the same and may affect the balanced growth of the primary, secondary and tertiary sectors of the economy. How can we deal with this situation in a manner which leads to a win-win outcome for all the stakeholders ?

The National Commission on Farmers, which I chaired during 2004-06, dealt with this question in detail and has also presented its recommendations in five reports submitted to the Ministry of Agriculture, Government of India. Based on a draft provided by NCF, the Union Minister of Agriculture and Food placed a National Policy for Farmers on the table of Parliament in November 2007. This is the first time either in colonial or independent India that a National Policy has been announced for farmers and not just for farming. There are numerous National Policies for Agriculture including the famous Royal Commission on Agriculture Report prepared during the colonial period, but none so far for farmers. The uniqueness of the National Policy of Farmers is that it calls for a paradigm shift from a purely tonnage- based approach to agricultural development to the socio-economic well-being of farm families. To quote: 'The aim of this Policy is to stimulate attitudes and actions which should result in assessing agricultural progress in terms of improvement in the income of farm families.' The Policy also calls for a major initiative for providing opportunities in an adequate measure for non-farm employment to rural families. The Policy Paper makes a commitment to launching a rural non-farm employment initiative jointly with all concerned agencies.

The Singur situation relating to land ownership underlines the urgency of approaching the problem of farming from two angles. First, the necessary public policy, technology, infrastructure and other support should be given to maximise the productivity and profitability of small and marginal farms. Second, steps should be taken to generate non-farm employment opportunities in the secondary and tertiary sectors through an integrated approach to agriculture, post-harvest technology and industry. China started its agricultural reform in the late 1970s with such a two-track approach. Steps were taken to help farmers to maximise the yield of major crops by providing irrigation and remunerative marketing opportunities. This is why the average yield of major crops like rice, wheat, etc., in China is more than double than that of our country. At the same time, China launched a well-planned Township and Village Enterprise Movement (TVE) in the early 1980s to shift about 100 million peasants to the non-farm employment sector. In China, land is socially owned and therefore such readjustment of income-earning opportunities could be done more easily than what is possible in our country. The results of this human-centered approach to livelihood security are summarised in a book titled *China's Township and Village Enterprises,* edited by He Kang, Former Minister of Agriculture of China and World Food Prize Laureate.[1]

Addressing a meeting in June 1987, Deng Xiaoping said,

> the greatest and most exceptional accomplishment in our rural reform has been the development of TVEs leading to the diversification of employment opportunities. The impact of TVEs has been dramatic and has helped China to provide income and work security for all in the rural areas.

The benefits of TVEs have been several, including a transition from unskilled to skilled work. They have proved to be China's unique road of industrialisation which has generated enormous manufacturing capacity in villages. This is why China has been able to attract extensive outsourcing of manufactured goods. He Kang has cited many examples of transformation of the rural economy through such an integrated approach to agriculture and industry.

The impact of China's twin strategy on livelihood security enhancement in rural areas has been striking in terms of human development indicators. For example, child malnutrition in India was about 52 per cent in 1992-93. It is currently about 46 per cent according to the National Family Health Surveys. In contrast, malnutrition of children in China came down steeply to 7 per cent during the same period. It is clear that all round human development takes place when income and work security becomes assured. This is why the NCF recommended that the State and Central governments should launch an integrated and dynamic non-farm employment initiative. NCF had also recommended that every State should develop a land-use plan earmarking land for different purposes like agriculture, industry, communication, housing and other human needs. It also proposed that where it becomes necessary to acquire farmland, the compensation given should be fair. For this purpose, NCF recommended the review and amendment of the prevailing Land Acquisition Act.

Taking the example of Singur, the land acquired for imparting an innovative and pro-low income orientation in the automobile industry is about 997 acres. This land belongs to about 13,000 farmers, of whom about 11,000 have received compensation. About 2,000 farmers owning nearly

---

[1] He Kang (2006). *China's Township and Village Enterprises.* Beijing: Foreign Languages Press.

300 acres of land are yet to accept the compensation. This situation exemplifies the position relating to the size of land holdings in our country. While a marginal farm is defined to be one hectare in size, most of the Singur farmers obviously belong to the sub-marginal category. Under such a situation, the need for multiple sources of livelihood becomes exceedingly important. This is why NCF suggested that agriculture and industry should prosper together and develop symbiotic linkages. Ultimately, about one-third of India's population may be able to have a reasonable income from the land. The rest will have to be engaged in the industrial, manufacturing and services sectors. This is also the road to ensuring that every citizen has a reasonable quality of life. I therefore support the view of the West Bengal Government that in addition to accelerated agricultural progress, triggered by the land reform measures introduced by the Government, there is need for creating greater avenues for employment in the industrial and services sectors.

Russi Lala in his book *The Romance of Tata Steel*[2] has described how the steel plant established by Jamsetji Tata over 100 years ago at Jamshedpur (the city was named after Jamsetji Tata much later) has transformed the entire economy and well-being of the people of that area. Jamsetji Tata established an ethical ground rule for the House of Tatas by emphasising that, 'we must give back to society many times more than what we have got from them,' Till today the House of Tatas has followed this philosophy, so that wherever their industrial units function, there is a marked improvement in all the human development indicators. Unfortunately, today land is being acquired by real estate developers at a very high cost for their own personal profit. In fact, many of the new entrants from India in the list of the world's richest persons are from the real estate business. Countries where land and material resources are overvalued and human resources undervalued tend to remain poor, as is happening in our country.

The Singur low-cost Nano car project is likely to confer multiple benefits to the community. As I have mentioned before, over 40 per cent of farmers interviewed during a survey expressed their desire to opt out of agriculture, if there is another option. In other words, the pressure of population on farmland is increasing and the younger generation is losing interest in agriculture. To retain youth in farming, it has to become both intellectually stimulating and

[2] Lala, Russi (2007). *The Romance of Tata Steel.* New Delhi: Penguin Books.

economically rewarding. Greater management efficiency is a downstream benefit of the way modern industry is organised and managed. Singur has an opportunity to become not only the Jamshedpur of West Bengal, but also the world capital for innovative, pro-low income, automobile technology and advanced small farm management methods.

In addition to the generous financial compensation provided by the State Government, an income security plan should be developed for the 11,000 farmers who have given land for the purpose of promoting the industrial renaissance of West Bengal. Children and the youth will suffer most if avenues for rural industrialisation get clogged. At least in their interest, short-term gains should not come in the way of the long-term prospect for a healthy and productive life for all. The younger generation will not forgive us if we do not open up new windows of opportunity for them for economic and intellectual fulfilment.

❑❑❑

Chapter 34

From Vision to Impact

During the last 21 years, the scientists and scholars of the M.S. Swaminathan Research Foundation (MSSRF) have been working on the design and implementation of projects which could have a large extrapolation domain in respect of imparting a pro-nature, pro-poor, pro-women and pro-sustainable livelihood orientation to technology development and dissemination. I would like to write about a few of the MSSRF initiatives, which have now become state, national and global programmes.

Mahila Kisan Sashaktikaran Pariyojana: Strengthening the role of women in agriculture

MSSRF initiated the Mahila Kisan Sashaktikaran Pariyojana in the Vidarbha region of Maharashtra in 2007 for empowering women farmers, including the widows of farmers who had committed suicide, in areas related to enhancing the productivity, profitability and sustainability of small-scale rain-fed farming. The empowerment measures incorporated access to technology, credit, inputs and market. Separately, an education programme was introduced for the children who had lost their fathers due to the agrarian crisis. Encouraged by the results of this small programme, Finance Minister

Shri Pranab Mukherji included funds in the Union Budget for 2010-11 for initiating a national Mahila Kisan Sashaktikaran Pariyojana. The Ministry of Rural Development, Government of India, which is in charge of administering this programme, has made it an integral part of its Rural Livelihood Mission. Recently, MSSRF was invited to undertake the Mahila Kisan programme in the Wardha and Yavatmal districts of Vidarbha from 2011 to 2014. This will include both technological and organisational empowerment. It is anticipated that by 2014, a well-organised Mahila Kisan Federation with a membership of over 3000 women farmers will emerge. There is a growing feminisation of agriculture in India, and it is hoped that the Wardha-Yavatmal Mahila Kisan Federation will be a forerunner to others at state and national level, capable of securing women farmers their entitlements. In addition to technology, inputs and market, women farmers also need services like crèches and day care centres. The gender-specific needs of mahila kisans, both as women and as farmers, will have to be met, if women are to play their rightful role in India's agricultural progress.

In addition to action at the grass-roots, MSSRF organised several consultations to prepare a draft Women Farmers' Entitlements Bill to be introduced in Parliament as a Private Member's Bill. The draft Bill is ready and is currently (2011) under circulation among women parliamentarians and gender specialists for their scrutiny and advice. It is hoped that this two pronged action — one at the village level, and the other, at the national policy level — will help the over 350 million women engaged in farming to contribute more effectively to agrarian prosperity and sustainable food security.

Pulses Villages: Bridging the demand-supply gap

To illustrate how the gap between demand and supply in pulses, which is one of the contributory factors to food inflation in the country, can be speedily bridged, MSSRF organised Pulses Villages in the Pudukottai and Ramanathapuram districts of Tamil Nadu over 15 years ago. In these Pulses Villages located in low rainfall areas, farmers undertook to harvest rainwater in farm ponds and cultivate pulses with appropriate varieties and soil fertility and agronomic management. Based on the success of this approach to accelerating progress in the production of pulses, a national programme for the establishment of Pulses Villages was recommended to the Union Finance Minister, who announced financial provision for starting 60,000

Pulses Villages in the country. A sum of Rs. 300 crore has been provided in the Union Budget for 2011-12 for organising 60,000 Pulses Villages. Already, the impact of this integrated and concentrated approach is becoming evident from the increase observed in pulses production from 14.66 million tonnes in 2009-10 to 16.51 million tonnes in 2010-11. Under the umbrella of the Pulses Village programme, special *Arhar* Villages (pigeon pea; *Cajanus cajan*) are being developed based on hybrid *arhar* strains. High-yielding *arhar* hybrids have been developed at the International Crops Research Institute for the Semi-arid Tropics (ICRISAT) located in Hyderabad. Women's Self-help Groups will be trained to become hybrid-seed producers and some of the pulses villages will be developed into Pulses Seed Villages for this purpose. This will enable the rapid spread of a yield revolution in pulses.

Nutri-cereals: Role in strengthening food security and climate-resilient farming

Almost from the early years of its establishment, MSSRF started working on underutilised or orphan crops such as a whole range of millets belonging to *Panicum, Pennisetum, Paspalum, Setaria, Eleucine* and other genera. These crops, normally classified as coarse cereals, are very nutritious and are rich both in macro- and micro nutrients. In fact, a combination of millet and *Moringa* (drumstick) provides most of the macro- and micro-nutrients needed by the body. The widespread hidden hunger now prevailing in the country as a result of a deficiency of iron, iodine, zinc, vitamin A, vitamin B_{12} and other needed micronutrients in the diet can be overcome at low cost through the consumption of millets and vegetables.

In 1992, MSSRF initiated in Kolli Hills in Tamil Nadu a programme for the revitalisation of culinary traditions involving a wide range of millets. A four-pronged strategy involving concurrent attention to conservation, cultivation, consumption and commerce was initiated. Commercialisation proved to be a trigger in the area of conservation, since farmers generally prefer to grow crops like rice, wheat or tapioca, for which there is a ready market. Similarly, in the Wayanad district of Kerala, tribal families were enabled to continue the conservation and consumption of tuber crops like *Dioscorea*. There is now a revival of interest in millets and other underutilised crops, both because of their ability to help in overcoming chronic and hidden hunger and their role in the design of climate-resilient farming systems.

In partnership with Bioversity International and the Agricultural Universities of Bangalore and Dharwar, and with financial support from the International Fund for Agricultural Development (IFAD) and the Swiss Agency for Development Cooperation (SDC), MSSRF has succeeded in introducing appropriate milling machines as well as markets for value-added products in a wide range of millets. Through several Policy Makers' Workshops and efforts in nutritional literacy, an understanding of the role of millets, tubers and other underutilised crops in improving rural nutrition and income in an era of climate change was promoted. Finance Minister Shri Pranab Mukherjee thus referred to *jowar* (sorghum), *bajra* (pearl millet), *ragi* (*Eleucine*) and minor millets as "nutri-cereals" and provided an allocation of Rs 300 crore in the Union Budget for 2011-12 for their popularisation.

In its draft National Food Security Bill, The National Advisory Council, headed by Shrimati Sonia Gandhi, has included millets among the staple grains that should be made available to food-insecure families, both in rural and urban India, at a highly concessional price through the public distribution system. If this Bill is approved and implemented, there will be a revival of interest in the cultivation and consumption of these nutrition-rich and climate-resilient crops. Agro-biodiversity hot spots can then become happy spots and will witness the dawn of an era of biohappiness where rural and tribal families are able to convert bioresources into jobs and income in an environmentally-sustainable and socially-equitable basis.

Another significant recent development is the initiation of a project on "Alleviating Poverty and Malnutrition in Agro-biodiversity Hotspots" with financial support from the Canadian International Food Security Research Fund (CIFSRF). The project is administered by the Canadian International Development Agency (CIDA) and the International Development Research Centre of Canada (IDRC) and involves partnerships with MSSRF, the University of Alberta, Canada, Bioversity International, the World Agroforestry Centre (ICRAF) and the World Food Programme (WFP). This five-year project (2011-16) will help to revitalise the *in situ* on-farm conservation traditions of tribal and rural families in the Kolli Hills area of Tamil Nadu, the Wayanad district of Kerala and the Koraput district of Orissa. MSSRF has been working with them for over 15 years. The contributions of the tribal families of Koraput have been recognised through the Equator Initiative Award at the UN Conference on Sustainable Development held at Johannesburg in 2002, and the Genome Savior Award by the Plant Variety Protection and Farmers'

Rights Authority of the Government of India in 2011. Thus, two decades of research and education carried out by MSSRF in the area of orphan crops have led to important research investment and public policy initiatives at the national and international level. The expansion of the food basket by increasing the number of crops which go into the daily diet will also impart stability to food security systems.

IDRC through CIFSRF is also supporting another project on strengthening rural food security through the production, processing and value-addition of nutritious millets. This project is being implemented in collaboration with McGill University, Canada and the University of Agricultural Sciences, Dharwad. MSSRF also coordinates the project activities assigned to the Himalayan Environmental Studies and Conservation Orgnisation (HESCO), Dehradun. This project capitalises on the progress earlier made by MSSRF in these crops with support received from the International Fund for Agricultural Development and Bioversity International.

Price Volatility and Hunger: Operation 2015

Nearly 70 per cent of the income of the poor goes to buy food. High prices therefore tend to reduce food intake by the poor, thus leading to the persistence of hunger. The extent of price volatility in recent years with reference to rice, wheat, maize and oil (petroleum products) is indicated in **Figure 1**.

The Agriculture Ministers of the G-20 Nations who met in Paris in June 2011 have emphasised that "small scale agricultural producers represent the majority of the food insecure in developing countries. Increasing their production and income would directly improve access to food among the most vulnerable and improve supply for local and domestic markets." The Ministers also decided to establish an Agricultural Market Information System, to start with in wheat, rice, maize and soybean, in order to improve agricultural market outlook and forecasts at the national and global levels.

MSSRF's work in this area has three major dimensions. The first is the development of village-level food security systems based on community Gene, Seed, Grain and Water Banks, which will help to store and distribute local nutritious grains like millets and pulses; the second encompasses the training of a cadre of "Community Hunger Fighters" who will be well versed in the science and art of overcoming both chronic and hidden hunger. The third dimension of MSSRF's work in the management of price volatility is a dynamic and location-specific market information system through Gyan Chaupals or Village Knowledge Centres. Many of these centres, now operating for over 15 years, provide timely information on the monsoon and the market. The behaviour of the monsoon and the market determines farmers' well-being. Hence, the Gyan Chaupals operated by local women and men give priority to empowering farm women and men with timely information on weather and market behaviour. Also, they provide information on food quality and safety, as well as on the entitlements of farm households to various government schemes.

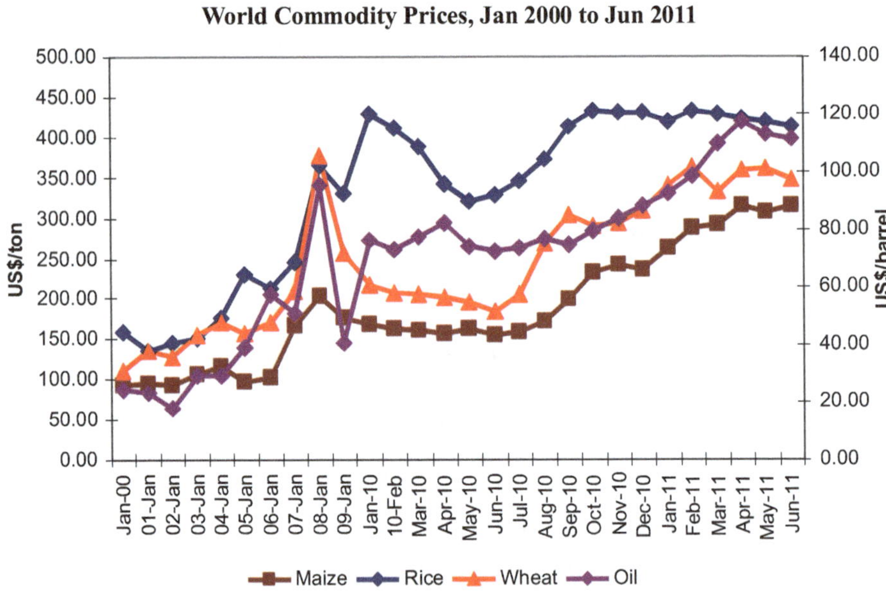

Figure 1

Source: FAO and US Energy Information Administration (data updated as on 29/06/2011)

The tribal areas where MSSRF is working in Tamil Nadu, Kerala and Odisha, as well as the Vidharba region of Maharashtra, are yet to achieve the progress necessary in the reduction of hunger and poverty to reach by 2015 the target set under the first among the UN Millennium Development Goals. Therefore, MSSRF in association with other partners has launched a programme titled "Operation 2015" to help these areas achieve UNMDG 1 by 2015. The programme consists of the following features:

- Adoption of a lifecycle approach in nutrition support programmes
- Promotion of a "deliver as one" method with reference to nutrition, clean drinking water, sanitation, environmental hygiene, and primary health care
- Payment of concurrent attention to small farm productivity improvement and producer-oriented marketing
- Encouragement of a food-cum-fortification approach (especially fortification of salt with iron and iodine) in respect of fighting chronic calorie deprivation and micronutrient deficiencies
- Establishment of a cadre (at least one woman and one man in every village) trained as Climate Risk Managers and Community Hunger Fighters

Thus, MSSRF hopes that the challenge of price volatility can be fought at the local community level as well as at national and global levels.

Seawater Farming

From 1990 onwards, MSSRF has been working on integrated coastal zone management, involving concurrent attention to the seaward and landward sides of the shoreline. The aim has been to strengthen both the ecological security of coastal areas and the livelihood security of coastal communities. A Coastal Systems Research (CSR) methodology was thus developed. The research activities included the conservation and restoration of mangrove wetlands, development of a Participatory Mangrove Forest Management System, generation of awareness of the importance of mangrove and non-mangrove bioshields in reducing the fury of coastal storms and tsunamis, and the breeding of salinity-tolerant rice, pulses and other crops of importance to coastal agriculture by transferring genes for salinity tolerance from mangrove species through marker-assisted selection of recombinant DNA technology. Eighteen years of sustained research in this field led to international patents being granted for the novel genetic combinations produced by MSSRF scientists for tolerance to abiotic stresses like salinity and drought. These include:

- US patent for the Dehydrin gene from *Avicennia marina* responsible for conferring salt tolerance in plants (Dr. Ajay Parida, Dr.Preeti Mehta and Dr. Gayatri Venkataraman)

- US patent for the Glutathione-S-transferase gene from *Prosopis juliflora* conferring drought tolerance in plants (Dr. Ajay Parida and Dr. Suja George)

Three more patents — for Phytosulfokine-Ü precursor sequence from *Avicienna marina* conferring stress tolerance, Antiporter gene from *Porteresia coarctata* conferring stress tolerance and Superoxidase dismutase gene for conferring abiotic stress tolerance in plants — have been filed and are in the process of being granted.

Marker-assisted breeding has resulted in developing location-specific transgenic lines in popular *indica* varieties (IR64, IR20, Ponni and ADT 43) showing 99.5 percent purity and enhanced salinity tolerance of 400mM of NaCl.

MSSRF's work led to the rehabilitation and replanting of 2400 ha of mangroves in Tamil Nadu, Andhra Pradesh and Odisha. The 2011 Coastal Regulation Zone Notification (6 January 2011) by the Government of India derives its scientific basis from MSSRF's research during the past 20 years and from two reports submitted by committees chaired by me.

On the basis of the projects proposed by MSSRF, both the Ministry of Environment and Forests (MoEF) and the Department of Science and Technology (DST) of the Government of India sanctioned funds for making effective use of seawater not only to raise bioshields, but also to initiate seawater farming projects involving integrated agro-forestry and mariculture techniques. The support from MoEF is through the Society of Integrated Coastal Management (SICOM). Seawater constitutes nearly 97 per cent of global water resources and Mahatma Gandhi rightly emphasised that it is a very important social resource. In 1930, Gandhiji's salt march was to manufacture salt in the Dandi beach in violation of the then prevailing government regulations. In the same year, C Rajagopalachari and Sardar Vedaratnam Pillai organised a salt satyagraha at Vedaranyam in Tamil Nadu. MSSRF organised a workshop at Vedaranyam on 26 December 2010 to highlight the need for undertaking the conversion of seawater into fresh water through halophytes possessing food and other economic value. The seawater farming project was included by DST under its WAR for Water Mission (Winning, Augmentation and Renovation). Steps have been initiated for establishing a genetic garden of halophytes in Vedaranyam, both to conserve the genetic resources of halophytes and to spread economically-attractive and environmentally-sustainable seawater farming methods. Under conditions of a potential rise in sea level, halophytes will become crops of the future in coastal areas.

Preserving Agricultural and Biodiversity Heritage Sites

During 2010-11, two important initiatives of MSSRF achieved wider impact. First, the Government of Tamil Nadu established genetic heritage gardens based on the description of ecosystems in the classical Sangam literature. These were set up at:

Kurinji (hill) – Yercaud, Salem District

Mullai (forest) – Sirumalai, Dindigul District

Marudham (wetland) – Maruthanallur, Kumbakonam, Thanjavur District

Neithal (coastal area) – Thirukadaiyur, Nagapattinam District

Palai (arid land) – Achadipirambu, Ramanathapuram District

In such genetic heritage gardens, the flora and fauna characteristic of each ecosystem will be preserved, which will help to spread the understanding of the value of such ecosystems. The garden in the Chennai campus of MSSRF also contains a replica of these five ecosystems described 2000 years ago.

The other important initiative relates to getting recognition for two Globally Important Agricultural Heritage Sites (GIAHS) under FAO's GIAHS programme. The project proposal seeking recognition for the Koraput rice genetic heritage site in Odisha has been prepared and forwarded to FAO. Here, tribal families have conserved a veritable mine of valuable genes in rice for hundreds of years. Recognition under FAO's GIAHS programme will help to give prestige to those conserving vanishing varieties and dying wisdom.

Another globally important agricultural heritage site is the Kuttanad area of Kerala where, for over a century, farmers have been practising farming below sea level. This system developed by farm families through practical experience involves the cultivation of rice during the monsoon season and fish during the non-rainy season. Unlike in the Netherlands, the Kuttanad farmers only put up low-cost temporary dykes. The GIAHS designation for the below sea level farming system developed by the farm families of Kuttanad will help to give recognition to the pioneers of this technology as well as refine it further. This will be particularly useful in the event of a rise in sea level as a result of global warming, as it now seems very likely. It is proposed to establish a Regional Training Centre for Below Sea Level Farming in Kuttanad, for the benefit of countries in this region — like the Maldives, Sri Lanka, Bangladesh and Thailand — which may have to undertake farming below sea level during this century.

Land and Water Care: Role of Global Soil Partnership

Since 2000, MSSRF, with financial support from the Tata Trusts and in association with the Punjab Agricultural University, Ludhiana and the Jawaharlal Nehru Krishi Vishwavidyalaya, Jabalpur, has been carrying out detailed studies on rainwater harvesting and efficient use, and watershed development and management. The emphasis in the current phase of this project is on maximising employment and income-generation opportunities for the watershed community through both on-farm and non-farm enterprises. The programme is hence known as "Bio-industrial Watershed" development. Small-scale market-linked enterprises supported by micro credit are promoted. Land-use decisions are also water-use decisions, and hence an integrated approach to land and water care is necessary to achieve an ever-green revolution leading to enhancement in productivity in perpetuity without associated ecological harm. Since land is a shrinking resource for agriculture and since there is a growing tendency to 'grab' prime farmland for non-farm purposes, such as for real estate and biofuel production, I proposed in October 2009, in my capacity as Chairman of the FAO's High Level External Committee (HLEC) on the UN Millennium Development Goals, the establishment of a Global Soil Partnership (GSP) for Food Security and Climate Change Adaptation and Mitigation. Both HLEC and the Director General of FAO have accepted this suggestion The Ministry of Environment and Forests has invited MSSRF to assist in developing strategies for sustainable food and nutrition security within the framework of a green economy. Obviously, a National Soil and Water Care programme involving all stakeholders, particularly farmers' associations, has to be an integral component of India's Rio +20 programme.

Human Resource Development

MSSRF's institution building philosophy has always been to concentrate on brains and not bricks. The sustained growth of MSSRF's Gyan Chaupal movement is a good example of the value of this approach. It is equally important that initiatives like Village Knowledge Centres are based on the principle of dynamic and location- specific information delivered in the local languages, based on a demand-driven approach. Local communities should also have a sense of ownership, as otherwise it will not be sustainable. The Jamsetji Tata National Virtual Academy, which now has nearly 1500 rural women and men as Fellows as well as 35 foreign Fellows, has become a valuable institutional device to build the self-esteem and capability of rural women and men belonging to socially- and economically-underprivileged families. In a recent review of the project, the reviewers concluded that the Academy has helped to convert ordinary people into extraordinary individuals.

It will be clear from the foregoing that the bottom line of the programmes undertaken by MSSRF during the last twenty years has been the well-being of rural and tribal families in an environmentally and socially sustainable manner. Unless we place faces before figures in our programmes dealing with human beings, we will not know whether the steps we have taken are really beneficial to those for whose welfare they are intended. "Remember your humanity" is therefore an effective method of monitoring the social impact of research and development programmes.

❑❑❑

Acronyms

AERB	Atomic Energy Regulatory Board
ARS	Agricultural Research Service
ASHA	Accredited Social Health Activist
BCKV	Bidhan Chandra Krishi Vishwavidyalaya
BPL	Below Poverty Line
CAMPA	Compensatory Afforestation Fund Management and Planning Authority
CIFSRF	Canadian International Food Security Research Fund
CAP	Common Agricultural Policy
CBD	Convention on Biological Diversity
CFS	Committee on Food Security
CIDA	Canadian International Development Agency
CIFE	Central Institute of Fisheries Education
CIMMYT	International Maize and Wheat Improvement Center
CRZ	Coastal Regulation Zone

CSC	Community Service Centre
CTBT	Comprehensive Test Ban Treaty
DST	Department of Science and Technology
FAO	Food and Agriculture Organization
FCI	Food Corporation of India
FMCT	Fissile Material Cut-off Treaty
GDP	Gross Domestic Product
GHG	Greenhouse Gases
GIAHS	Globally Important Agricultural Heritage Sites
GNH	Gross National Happiness
GM	Genetically Modified
GSP	Global Soil Partnership
GWP	Global Water Partnership
HESCO	Himalayan Environmental Studies and Conservation Organisation
HEU	Highly Enriched Uranium
HLEC	High Level External Committee
IADP	Intensive Agriculture District Programme
IAEA	International Atomic Energy Agency
ICAR	Indian Council of Agricultural Research
ICDS	Integrated Child Development Services
ICMR	Indian Council of Medical Research
ICRISAT	International Crops Research Institute for the Semi-Arid Tropics
ICT	Information and Communication Technologies

IDRC	International Development Research Centre of Canada
IFAD	International Fund for Agricultural Development
IMR	Infant Mortality Rate
INCOIS	Indian National Centre for Ocean Information Services
ISME	International Society of Mangrove Ecosystems
ISRO	Indian Space Research Organisation
IUCN	International Union for the Conservation of Nature and Natural Resources
IWMI	International Water Management Institute
LBW	Low Birthweight
LEISA	Low External Input Sustainable Aquaculture
MFA	Most Favourable Area
MGNREGA	Mahatma Gandhi National Rural Employment Guarantee Programme
MoEF	Ministry of Environment and Forests
MSA	Most Seriously-Affected Area
MSP	Minimum Support Price
MSSRF	M.S. Swaminathan Research Foundation
NABARD	National Bank for Agriculture and Rural Development
NAC	National Advisory Council
NASSCOM	National Association of Software and Services Companies
NBDB	National Bioresource Development Board
NCF	National Commission on Farmers
NFHS	National Family Health Survey
NPP	Net Primary Productivity

NPT	Nuclear Non-Proliferation Treaty
NREGA	National Rural Employment Guarantee Act
NSSO	National Sample Survey Organisation
NVA	National Virtual Academy
PDS	Public Distribution System
RET	Rare, Endangered and Threatened
SAZ	Special Agricultural Zone
SEZ	Special Economic Zone
SFAC	Small Farmers' Agri-business Consortium
SHG	Self-Help Group
SICOM	Society of Integrated Coastal Management
SRI	System of Rice Intensification
SDC	Swiss Agency for Development Cooperation
TINP	Tamil Nadu Integrated Nutrition Project
TVE	Township and Village Enterprise
UNESCO	UN Educational, Scientific and Cultural Organization
UNFPA	UN Population Fund
UNMDG	UN Millennium Development Goals
USAID	United States Agency for International Development
VKC	Village Knowledge Centre
VRC	Village Resource Centre
WFP	World Food Programme
WTO	World Trade Organization

❑❑❑

Zeitfracht Medien GmbH
Ferdinand-Jühlke-Straße 7
99095 Erfurt, Deutschland
produktsicherheit@kolibri360.de